EUROPEAN CLOCKS

Frontispiece: A small Black Forest Alarm clock, with painted dial, *c.* 1840. Wooden frame, weights supported by cords instead of chains.

EUROPEAN

CLOCKS

E. J. Tyler

WARD LOCK & COMPANY LIMITED

London and Sydney

To my wife

Made and printed
in Great Britain by
William Clowes and Sons, Limited
London and Beccles

Set in Monotype 11/12 pt. Perpetua

CONTENTS

PLATE 1   An astrolabe for telling the time by the stars. The
instrument shown is that used in the 16th-century.

I

THE EARLY HISTORY OF THE MECHANICAL CLOCK

The mechanical clock first appears in written records towards the end of the thirteenth century, but even here the historian has to tread warily, for the text employs the word 'Horologium' or derivatives of it, which can also mean a sundial, water clock or some other form of timekeeper. We can safely assume, however, that the mechanical clock was well established before 1350 or so for Giovanni di Dondi of Padua prepared an elaborate treatise in Latin describing a Planetarium, indicating the motion of seven planets, which he constructed himself during the years 1346–64, and stated that the driving mechanism was an ordinary clock which it was not necessary to describe as the reader would be familiar with it. Fourteenth-century writings tend to accept the clock as commonplace unless it had functions other than the mere recording of the hours. Surely some writers must have recorded the invention and if they did why have all their manuscripts been lost?

The mechanical clock consists in essentials of four parts:

1 Motive power.
2 Train of wheels to transmit the motive power.
3 Escapement for controlling the speed at which the machine runs.
4 Means for recording the motion of the machine either continuously by a hand on a dial, or intermittently by ringing a

bell at a predetermined time as on an alarm clock, or regularly by announcing the hours by a definite number of blows on a bell.

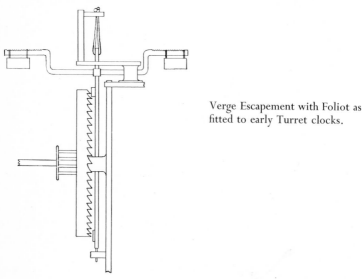

Verge Escapement with Foliot as fitted to early Turret clocks.

While the dial will indicate the time by means of the main clock mechanism, the sounding of an alarm or of a certain number of blows requires extra mechanism to accomplish it. It is possible to strike one blow at the hour or half hour using the main mechanism only without requiring an extra train, and an alarm can be sounded by arranging for the going part to release a bell which has been held to one side and allow it to ring until it comes to rest. For all practical purposes, however, it is necessary to incorporate extra mechanism when the clock is desired to strike or sound an alarm.

By the late thirteenth century all the essential mechanism was in existence except the escapement, but we shall probably never know the name of the genius who invented it and combined it with the other essentials to produce a clock. It may have been an armourer who accompanied the Crusaders to the Holy Land and saw some such device being used by the Arabs for a purpose entirely unconnected with time measurement. The idea for motive power was given by the simple winch for raising a bucket from a well, toothed wheels were already known in windmills, and the sundial would suggest a means of recording the machine's motion.

It should not be forgotten that the earliest mechanical clocks did not show a great improvement over other forms of timekeeper. They were controlled by a foliot balance which had no natural timekeeping period of its own, and checked the speed at which the mechanism ran simply by its own inertia. As the oil thickened, the speed of the mechanism would slowly reduce. Daily comparison with a sundial was a necessity.

It should not be forgotten either that the clock measures hours of equal length, while previous systems had divided day and night into the same number of hours, which would vary in length with the seasons. Some opposition to the idea of hours of constant length must have been encountered at first, but soon after the invention of the mechanical clock the various countries of Europe adopted the new system. They did not all immediately adopt the same system of division, however.

The monasteries had previously divided the day into twelve hours from sunrise to sunset and the English system divides day and night into twelve hours each but counts twice twelve beginning at noon and midnight. The Italians divided the day into twenty four hours beginning at sunset, and early Italian clocks were arranged to strike from one to twenty four. This system continued after most other countries had adopted the 2×12 system.

In Nuremburg the day and night were divided into eight to sixteen hours according to the varying seasons and there is a clock in the Germanisches National Museum in that city which has its dial arranged to accommodate the changing number of hours.

It has been suggested that the earliest clocks were created simply for driving working models of the planetary system such as that of Dondi mentioned previously. In such cases an error of an hour or so in a day was not of great importance and no one worried very much, but when the clock began to be used to measure the passage of time during a day, the point would become more pressing. The clock as an independent mechanism was first used to give the times of services in monasteries. This would probably have involved a device for ringing a bell at predetermined intervals and would therefore have formed a primitive alarm clock to draw the Sacristan's attention to the time, and remind him to summon the monks to prayer, probably by ringing a larger bell in a tower.

There are two schools of thought in Horological history. One believes that the large turret clock movement came first while the other believes that the small clock came first and that the turret clock was an enlarged version of it. Whichever opinion is correct, the monastery clock would have resulted in bells being rung that could be heard outside the monastery. The lay public would have found those bells useful and there would have been an incentive for towns to possess clocks of their own so that the sound of bells could be carried over a wider area. Ungerer gives a list of towns together with the first known date on which they possessed a clock and it can be seen that many of the dates are quite early. Milan claims to have had the first one in 1335.

The word 'clock' is connected with the French *cloche* and the German *Glocke*, both meaning 'bell', and the importance of early clocks was to announce the time by striking a bell at intervals, either continuously as an alarm or by sounding a definite number of blows. The latter function needs more complicated mechanism than the former and embodies the following essentials:

1 Motive power.
2 Wheel containing pins or cams to raise the hammer tail and let it fall again.
3 Connection from the going train to release the striking work as required.
4 A device to determine the number of blows struck and lock the train after this number has been given.

Striking mechanism

5 A speed regulator, usually in the form of a fan or fly.

The release of the striking work is usually effected by means of a pin in the rim of a wheel in the clock which makes one revolution per hour. This pin raises a lever as the wheel rotates, and, in the oldest type of striking work, the arm which is holding the striking train locked is fixed to the same centre as this lever and is raised at the same time. When the levers have been raised to a certain height, the train is released and the clock strikes. As it is necessary to lock the train after a required number of blows has been sounded, the levers must be capable of descending when this point has been reached, but if no special arrangements were made, the pin which actuated the release in the first place would prevent

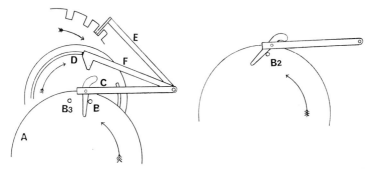

Striking without Warning as used on the Salisbury clock. Wheel A of the going train carries Pin B which acts on the hinged end of the Lifting Piece C, pushing it to one side until its opposite end makes contact with the Lifting Piece itself as in the right-hand sketch, B2. The three arms C, E, F all move together. Further movement of the pin causes the Lifting Piece system to rise, and as soon as the middle lever F has been raised sufficiently for the Hoop D on the second wheel of the Striking Train to press against its sloping edge, the Striking Train begins to run. As the Hoop passes under the sloping-edge it raises the lever system still higher, and the hinged end falls off the pin which raised it into the position it occupied before it was moved. B3. By the time the break in the hoop has come round again, the top arm, E, of the system is riding on the edge of the Locking Plate which slowly rotates until a notch comes below the arm. The lever system is then allowed to fall, and the train is locked by the edge of the hoop butting against the middle Lifting Piece F again.

this. In the Salisbury clock of 1386 this obstacle was overcome by having the lifting pin act on a jointed piece at the end of the lever rather than the lever itself. As the striking train begins to run, it raises the lever system still higher and allows the jointed piece to fall away so that when it descends it does so on the other side of the pin which raised it. A similar system of striking was used in Dutch clocks until a comparatively late date, but the jointed piece was moved by a spring instead of by gravity.

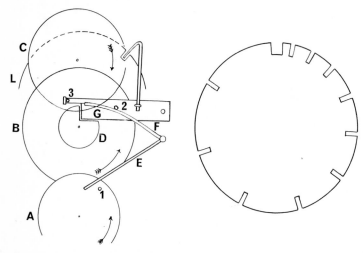

Striking Work with Warning as fitted in a normal Black Forest Clock   Wheel A is part of the motion work and raises Lifting Piece E by pin 1. E carries another arm F which raises pin 2 on Lifting Piece G. The train is locked by pin 3 on wheel C butting against the end of Lifting Piece G. When G has been raised enough to release pin 3, the wheel C makes almost one revolution until pin 3 is caught by the end of arm F, and Cam D on wheel B turns slightly so that G cannot descent to its former position. When E drops off pin 1, both Lifting Pieces descend, G being caught by Cam D on wheel B which rotates once for every blow struck. Only when the upper arm of G, which rides on the Locking Plate L, can enter a slot, is G able to descend to the full depth of the notch in the Cam, and the train is locked once more by pin 3 being caught by the arm of G.

The right-hand sketch shows a complete Locking Plate.

Control of striking mechanism

The later method of strike release employs two lever systems instead of one. The first lever is operated by the going part as before, and after it has been raised a certain distance a second arm attached to it begins to raise the other lever system whose function is to keep the train locked. When this has been raised to a certain height the train begins to run but is immediately stopped again by another arm on the first lever system interrupting one of the more rapidly moving wheels at the top of the striking train. This is known as the 'Warning'. The striking work is finally released when the first lever drops off the pin that raised it and it continues to run until the control mechanism locks it again. The advantage of this system is that it is easy to arrange for the lever to drop off its pin when the hand is exactly on the hour, while with the previous system the time of striking may vary by a minute or two every time it is released. 'Warning' was probably invented about 1390, for the Salisbury clock of 1386 does not possess it while the Wells one of 1392 does.

Controlling the number of blows struck was originally performed by means of a count wheel or locking plate, a wheel which had a number of notches cut in its edge, the distance between successive notches increasing so that every time the train runs one more blow may be struck. The function of the locking plate is to keep the lever system raised and only allow it to fall, and thereby lock the train, when the arm riding on the edge of the locking plate reaches a notch.

It should be noted that the lever falling into the notch does not constitute the actual locking, which is performed by another arm of the lever butting against a pin or the edge of a three-quarter hoop on a more rapidly moving wheel in the train. Locking against the edges of the locking plate itself would cause enough friction to prevent the going side of the clock from unlocking the striking train.

The locking plate is the form of striking control used on the oldest clocks and is still used today for cuckoo clocks. It has the serious disadvantage that if the train is accidentally released the

number of blows struck does not correspond with the time indicated by the hands. To overcome this difficulty the rack-striking
mechanism was invented about 1676. In this method, the number
of blows struck is controlled by a spirally shaped piece of metal
which is carried round with the hour hand, and on account of its

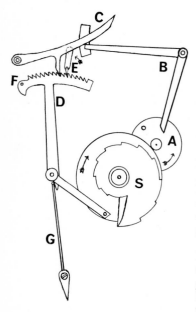

Rack striking Mechanism. The clock is
in the process of striking twelve and
the Gathering Pallet is about to move
the sixth tooth of the Rack while the
sixth blow is being struck The snail
S rotates with the hour hand. Wheel
A rotates once per hour.

The striking work is released by
the pin A raising Lifting Piece B
which in turn raises the Detent C
which is holding the Rack D in
position. When the Detent is clear
of the Rack, the latter falls, the
distance being regulated by the Snail.
The train, which till now has been
held by the tail of the Gathering Pallet
E butting against the pin F in the
Rack, begins to run. Almost immediately it is held up again by the
end of B, which projects through the
plate, arresting a pin on the so called
'Warning Wheel' of the train. When
B falls off the pin which has raised it,
the train is free to run and the clock
goes on striking until the Rack has
been completely gathered up and the
tail of the Gathering Pallet butts
against Pin F once more. G is a
spring to help the rack fall smartly.

shape is called the 'snail'. When the clock strikes, a toothed rack,
which consists of a portion of a large toothed wheel, is allowed to
fall on the snail, the position of the snail determining the distance

the rack falls. As the train runs, the rack is gathered up one tooth for every blow struck, hence the number of blows is controlled by the position of the snail. It is possible to make the clock repeat an infinite number of times, and clocks in the past were often furnished with a string that could be pulled during the night to indicate the time; a very useful feature in the days when matches or electric light did not exist.

The French *Comtoise* clocks (see Chapter 3) are often arranged to strike the hour again two minutes after having done so and employ rack striking with a slight modification of the release mechanism to do this. On these clocks the rack drops vertically on to the snail and is straight instead of forming a segment of a wheel.

The early town clocks were almost certainly turret clocks and special towers would have been needed to house them. The Town Hall would be an obvious place for the clock, but in some cases separate towers would be erected because of geographical or other considerations. The clocks would have been made by itinerant workmen who worked much as the itinerant bell-founders. These men would also be called upon to do repairs to existing clocks when necessary. The possession of a clock endowed great prestige on a town and not only on a town but also on a church if it could afford to possess one. Along with the towns we therefore find the churches and cathedrals installing clocks at this period. Some of these clocks did more than just tell the time and strike. The function of indicating the position of heavenly bodies assumed great importance and also automatic calendars showing the important days in the Church's year were incorporated. Moving figures to pass in procession or strike the hours were provided and often a cock which crew and flapped its wings would surmount the clock.

In such a machine as this the timekeeping function was of lesser importance, and the clock would therefore be placed inside the building for protection rather than in a tower where its chief use would be to sound the hours and give the public the time. Its position would not preclude its being arranged to ring bells in a tower also if desired, but its main function was to be inspected by visitors who would no doubt contribute financially towards the extremely high cost of the clock. Complicated dials would rapidly deteriorate if exposed to the weather. The use of an exterior dial on a tower came later. The oldest illustrations of exterior dials that have come to light are of the old church at Amsterdam dated 1544 showing four dials, and a general view of Innsbruck dated 1577 showing dials on two different buildings. Artistic evidence of exterior dials prior to this is given by a panel in the Pinto collection of wooden bygones showing a tower with a dial that revolves past a stationary hand. This is entitled 'The Annunciation' and is by Fr Damiano de Bergomo and dated 1536. A picture in the Louvre by Maitre Rhénan, 'Pieta de St Germain des Prés' includes a suggestion of an exterior dial on a tower and dates from about 1500. The picture 'Virgin with Angels' by Jan Gossaert 1472–1536 in Palermo museum, also appears to show a clock dial in the background.

Documentary evidence of early exterior dials is scanty. In 1344 the Dean and Chapter of St Paul's Cathedral, London, made a contract with Walter the Orgoner of Southwark to supply and fix a dial for the clock which may have been an exterior one as a roof is mentioned. The handbook of the National Museum of Wales quotes a poem written betwen 1340 and 1370 which contains the line 'Woe to the black faced clock which woke me'. 'Black faced' may refer to a black dial or may simply be a poetical term of abuse. Dr C. F. C. Beeson in his study of clocks in Oxfordshire has discovered a reference to an exterior dial in Oxford in 1505.

The importance of the clock in the early days was telling the time by sound, and it was not long before the utility of ringing a bell or bells more frequently than once per hour was appreciated. The earliest mechanism for sounding the quarters that is known is that of the great clock at Rouen dated 1389. However useful quarter chimes may have been, the extra expense involved must have delayed their general adoption. Clocks in those days were extremely dear. Not only was very little skilled labour available to make them, but the making of a clock took a very long time owing to the lack of technological facilities.

Quarter chimes led the way to the carillon which is believed to have been invented at Aalst in the Low Countries late in the fifteenth century. Instead of the clock sounding the quarters in the usual manner, it would play a portion of the tune set up on its drum, giving a longer selection at half past and quarter to and possibly playing the whole tune before the hour was sounded. Sometimes it would be arranged to sound a few notes seven and a half minutes after each quarter. Many towns employed a carillonist whose duty was to play tunes on the bells with a keyboard and also alter the tunes set up on the drum of the clock as required. The carillon was especially popular in the country of its origin, but it did not spread to Britain to any great extent.

The monastery type of alarm clock was probably put to secular as well as monastic use. A town which could not afford a large striking turret clock could instal one of these, and employ an odd job man whose other duties might have included sweeping the market square to pay attention to the ringing of the alarm and then ring a bell in a tower to advise the townspeople of the time. As striking clocks became more popular his function would cease, but we find many instances of wooden effigies erected on clocks for the purpose of striking bells as a reminder of the former human agent. These effigies are known as Jacks and appear both in Britain and Europe. The name may be derived from the Latin 'Jaccomarchiadus' meaning 'a man in a suit of armour' or perhaps simply from the word 'jack' meaning a piece of apparatus, e.g. 'boot jack', 'roasting jack' etc. Their popularity is such that they have been made from mediaeval times to the present, and it is often difficult to date a given example accurately.

There are not many very old clocks preserved on the Continent. That of Rouen dated 1389 is probably the oldest, but this has undergone some alteration during its existence. The oldest known clock in England is considered to be that of Salisbury

Cathedral. It was discovered in 1929 by Mr T. R. Robinson who noticed a number of similarities to the Wells clock, which had been preserved in the Science Museum for some years. In 1956 the Salisbury clock was restored as nearly as possible to its original condition with a verge escapement and foliot balance and is now exhibited going in the nave of the cathedral. The additional parts made for it have been distinctively coloured, and probably quite a large proportion of the remaining mechanism is original. It can be considered to illustrate the appearance of a typical clock of the middle ages. It never had a dial and strikes the hours only.

The estimated date of the Salisbury clock is 1386 and that of Wells 1392. The similarities in the movements of the two clocks make it highly probable that they were made by the same workmen, but Wells shows a much higher state of development. Wells chimes the quarters and also incorporates 'warning' in its striking which Salisbury does not. For a long time there was a legend current that the four West of England clocks, i.e. Wells, Wimborne, Ottery St Mary and Exeter had been made by a monk of Glastonbury called Peter Lightfoot. This is now generally discredited but is still believed in certain circles. Out of these clocks, Wells is fourteenth century, Ottery St Mary possibly sixteenth century, Exeter is a mixture of styles although the quarter chiming part may be old, and Wimborne is eighteenth century. The Lightfoot legend never seems to have been attached to the Salisbury clock.

PLATE 2 The verge escapement of a Gothic Chamber clock.
The wheels and pinions would have been filed by hand.

PLATE 3   A large turret clock movement from the St. Jacobs-toren, The Hague, 1544. The foliot can be seen above the movement, top right. The centre drum is for the carillon and the trip pieces are adjustable so that different tunes can be played.

2

THE BEGINNING OF THE DOMESTIC CLOCK

After the clock had become established as a feature of Town Halls and large churches, the smaller churches, especially those in small towns and villages would also desire clocks but not have the means to pay for ones similar to the large and elaborate machines then in use. A demand would have been created for a smaller simpler turret clock which began to be satisfied in the fifteenth century. Such examples that have come down to us are mostly of the sixteenth or seventeenth centuries, but certain evidence, particularly from a manuscript in the Bibliotheque Royale at Brussels, *L'Horloge de Sapience*, shows that smaller clocks were indeed being made in the fifteenth century. After the erection of clocks in public buildings it could be expected that the rich noblemen would want to possess clocks for their castles, not only in the form of turret clocks but also smaller clocks for their own private apartments. During the fifteenth century, therefore, we find scaled down examples of turret clock movements being made for domestic use. As the full size clocks were contained in towers, the frame of the domestic version would be constructed with buttresses to resemble a tower, and the bell for striking would be placed on top with a rudimentary form of spire to support it. The shape of a turret clock movement tended to approach a cube, but the domestic clock, being modelled on a tower, was much taller.

Early domestic clocks

A new technique was employed in making these clocks. While the large turret clocks were made by beating hot iron on an anvil and afterwards filing the parts to shape, i.e. using the craft of the blacksmith, the smaller clocks were made mostly by filing the metal cold with the minimum of blacksmithing, and employing techniques more akin to the locksmith's craft. The earliest makers of domestic clocks were probably therefore locksmiths. The earliest makers of watches followed this trade.

The domestic clock was weight driven like its larger counterpart and showed no great difference in its mechanism. It was usual to add an extra wheel to each train to enable the clock to run with a proportionally shorter fall for the weight so that the clock could hang at a convenient height for its dial to be seen. In the case of turret clocks, the weights would run down a tower and it was not of great importance to limit their fall. Even so, the early domestic clocks would need winding every twelve to fifteen hours. Such a duty would be performed only by the master of the house himself or a very responsible official of the household and the clock would be frequently checked against a sundial. Where the clock was fitted with a foliot balance, it could be regulated after a fashion by moving the small weights on the foliot nearer to or farther from its centre, but where a wheel balance was provided the only means of regulation would be to add small weights to the driving weight or remove them as necessary. The accuracy of the clock would have shown no improvement over that of its larger predecessor.

The weight driven chamber clock appeared about the end of the fourteenth century and was made with little modification until the mid-seventeenth. France, Germany and Italy produced the best known examples, and other countries such as Switzerland also made these clocks, but not in such great numbers. In particular, the Liechti family of Winterthur became famous for this type of clock, and the products of several generations of the family are known. Some of these early clocks were fitted with alarms and would therefore be useful to arouse their owners from sleep. The clocks were not readily portable and would therefore have to be hung in the bedchamber, which limited their usefulness during the day. By this time the fashion of sleeping in a separate room would have established itself, and a portable timekeeper would therefore have been extremely useful. The main difficulty was the weight drive.

The problem was solved by using a steel coiled spring as motive power instead of a weight. It was formerly believed that the spring came into use about 1500, but recent researches by Mr P. G. Coole of the British Museum have indicated that it may have been used as early as 1407. There were two main difficulties in the application of the mainspring. The first was the making of the spring itself, as in those days steel was only capable of being produced in small quantities and the quality could not be guaranteed. A homogenous piece of steel was an impossibility at that time, and several springs might have to be made and discarded before one could be produced that would do its job in a satisfactory manner. The second difficulty was that a spring exerts a greater force when it is wound up than when it is nearly run down, and with the verge escapement this is fatal for timekeeping. It was necessary, therefore, to provide some device to equalise the pull of the spring to get as nearly as possible a constant driving force. This was done by means of a fusee.

A fusee consists of a tapered piece of metal, circular in section with a spiral groove cut on its surface. The mainspring is contained in a drum called a barrel, and a gut line is wound round the outside of the barrel and connected to the fusee at its largest diameter. The clock is wound by rotating the fusee and drawing the line on to it and at the same time winding it off the barrel. The rotation of the barrel winds the spring. As the last turn of line comes off the barrel, it moves a stop device which prevents further winding and the possible breakage of the line. When the clock is fully wound, the line is nearly all on the fusee, and the force of the spring pulls on the line, which acts on the fusee at its smallest diameter. As the spring gets weaker, the pull is transmitted to the fusee on an ever increasing diameter, giving an easier leverage, and in practice a nearly constant force is obtained. It is arranged that when all the line has returned to the barrel there is still a little tension on it to prevent its coming slack and possibly jumping the grooves of the fusee at the next winding.

The Brussels manuscript previously mentioned shows a movement on a table consisting of two plates held apart by pillars, with a fusee and barrel between them. This was the form that spring clocks eventually took, but a portrait of a Burgundian nobleman in the Museum of Fine Arts at Antwerp has a representation of a clock in the corner that looks externally like the usual weight driven chamber clock, but which on closer inspection proves to be a spring driven clock with the barrels containing the springs hidden in the base. Other clocks have been seen in museums with the springs arranged like this, but none have been discovered so far that present such a close likeness to a weight driven clock as the one in the picture. It is highly probable that the artist was working from a real clock and not using his imagination when he painted the portrait, as the nobles of that time liked to have their treasured possessions included in their portraits. It will be interesting to see if any further clocks of this type come to light. The clock supposed to have belonged to Philip the Good, Duke of Burgundy, about 1430 and now preserved in the Germanisches Museum at Nuremburg shows a certain similarity to this clock, but as grave doubts have been cast on the authenticity of the Philip the Good clock, one should be reluctant to draw any inferences from the similarity.

The spring driven clock was intended to stand on a table, and as it would be nearer to the person who wished to refer to it it was possible to make the dial smaller than before. As time went on, smaller clocks began to be made, and by the early sixteenth century they were small enough to be carried on the person, and so the watch came into being. The subsequent history of the watch is complex enough to need a book of its own, and several writers have already dealt with this theme.

As the movement became smaller, the fusee had to be smaller also, and so became more difficult to make. An alternative in the form of a brake, known as a *Stackfreed*, was provided in Germany. It was used mostly for watches, although some table clocks have it, and it was abandoned in the early seventeenth century. Other countries seem to have ignored it.

Table clocks could be divided broadly into two groups, those with vertical and those with horizontal dials. The horizontal dial group developed into the early watches. As the sixteenth century progressed, table clocks acquired astronomical and other subsidiary dials. France, Germany and Italy supplied the bulk of these clocks and examples from other countries are rare. Augsburg and Nuremburg became celebrated for their clocks and the former city was especially noted for its automata.

The table clock marked a definite break with tradition in that the mechanism became boxed in, i.e. the clock had a case. The early wall clocks consisted simply of the mechanism with a little decoration, and the movement of the wheels could be seen. The early spring clock in the portrait mentioned previously is constructed on the same lines as the weight driven clocks, but the scarcity of spring driven clocks of this type probably shows that this type had shortcomings which the craftsmen of the day sought to eradicate by adopting the enclosed movement.

It was in the sixteenth century that brass began to be used for clock movements. Di Dondi describes his clock as being made entirely of brass and bronze, but it was a good two hundred years before brass was used again by clockmakers. Di Dondi was a scientist and not a tradesman and therefore he would use what he thought best. The clockmakers were the technological descendants of either blacksmiths or locksmiths and would go on using the metal that they were used to, i.e. iron. They may even have doubted the ability of brass to stand up to the constant wear.

Screws made their appearance about 1560. Previously the parts of the mechanism would have been fastened together by means of wedges, but as screwing tackle became more widely used the convenience of screw fastenings over the previous method would have become obvious. Early dies forced the threads on the screws rather than cut them, and no workman would have made any more thread than was absolutely necessary for the job in hand. No screws were standardised and every watchmaker was a law unto himself as far as screws were concerned.

More skilled labour was available on the Continent than in Britain during the middle ages. As far as can be ascertained there were no British craftsmen capable of making a clock until the late sixteenth century. If any church or town in this country desired a clock, the workmen had to come from abroad. Edward III granted a charter of protection to three 'Orologiers' in 1368, one of them at least coming from Delft, and they were to be allowed to exercise their craft in the realm. They probably worked on clocks for the king at Westminster, Queenborough, and Langley about this period, and the clocks at Salisbury and Wells have also been attributed to them. Three nameless Lombards were working on a clock for the great tower of Windsor Castle in 1352.

Repairers of clocks were known in Britain from the middle ages and references to them all seem to indicate that they were working on turret clocks, but it is doubtful if any of them were English. For instance there is a reference to 'Roger the Clock-maker' being sent from Barnstaple when Exeter Cathedral clock needed repair in 1424, and the 'clokke maker of Kolcester' (Colchester), repaired a clock for the Duke of Norfolk in 1483. The word 'make' was formerly used in the sense of 'mend', so it cannot necessarily be assumed that these 'clockmakers' actually made clocks or did anything else than repair or maintain existing clocks. When Nonsuch Palace was being built at the end of Henry VIII's reign we have records of clockmakers employed there but it is doubtful if any of them were English. 'A frenchman' repaired the church clock at Rye in Sussex in 1533, and Lewis Billiard, a native of Gascony, supplied a new clock for that church in 1561, which is still basically the movement in use there today. Henry VIII possessed a number of clocks and watches, all of which must have been imported from the Continent or made by visiting workmen. Elizabeth I also possessed many clocks and watches and had her own clockmaker in the person of Nicholas Urseau who was of French descent, but her clock keeper was Bartholomew Newsam who also received the office of clockmaker in 1590 on the death of Urseau. He thus became the first English Royal Clockmaker. Newsam is believed to have been a Yorkshireman and a charming little vertical table clock of his making is preserved in the British Museum.

Mention should also be made of Randolph Bull who made the clockwork for Thomas Dallam's organ which was presented by Queen Elizabeth I to the Sultan of Turkey in 1599. Bull later became Royal Clockmaker. Other names of British clockmakers begin to appear about this time, and at last the craft established itself in this country.

As soon as the making of clocks began in Britain, a British style emerged. This is popularly known as the 'Lantern' clock, but is also known as 'Cromwellian' or 'Bedpost' and by various other names. Late examples in which the dial is much wider than the movement are called 'Sheepshead'. The movement of these clocks was basically the weight driven wall clock of the Continent, but most of the metal used was brass and the clock was generally not so high. A wheel balance was always used. So far, no evidence has been forthcoming of an English Lantern clock with a foliot. The design would not allow for this, as the bell was not very far from the top plate of the movement, and was flanked on three sides by frets which would have rendered the adjustment of the small weights difficult. Regulation of a Lantern clock would always have been carried out by increasing or decreasing the amount of lead shot carried in a hollow on top of the going weight, and thereby altering the amount of driving force available.

The Lantern clock had a very long life. It began in the reign of Elizabeth I and continued to be made until that of George III, although by this time it was only being made in the country. Later examples are pendulum controlled and have a single driving weight which provides power for both going and striking trains. The earliest clocks possessed a separate weight for each train, and these were arranged to hang on opposite sides of the clock. The first wheel of each train would be provided with clickwork to allow separate winding, and they would rotate in opposite directions. When the single weight was introduced after pendulum control had been applied, both wheels would rotate in the same direction and clickwork would be applied to the striking side only, ensuring that the clock would continue to go while it was being wound. Such a refinement would have been meaningless on the old balance clocks with their notoriously erratic timekeeping.

The Lantern clock at first had a narrow chapter ring with stumpy figures. The hour hand was of sturdy construction to permit of its being set to time, and the inner edge of the chapter ring was engraved with quarter hour marks to allow a closer approximation to the time to be read. As the seventeenth century progressed, the dials became larger and the figures longer, and after the invention of the pendulum, minute hands were added. Smaller versions with an alarm only and no striking work were produced for travelling purposes, neatly packed into a travelling case with their weights ready to be hung on the wall of a bedroom in an inn to wake the owner in time for his next day's journey. The popularity of the style lasted until long after they had ceased to be made as a regular item of clockmaker's wares. During the nineteenth century many old clocks had their movements replaced by contemporary spring driven ones and were adapted for standing on a mantelpiece which is quite out of keeping with this type. Later still, small versions with platform escapements were sold for use on desks or bedside tables.

The frets flanking the bell are an important feature in the design of a Lantern clock. The dolphin pattern was always popular and various arrangements of foliage, armorial bearings and even architectural motifs are met with that can often give a clue to a maker, a period or a particular location.

The Clockmakers' Company

Once the making of clocks was established in London, the makers felt a need for joint action to protect their craft. Any clocks produced in Britain before about 1600 had been made by foreigners, but as the native talent was developed a desire arose to eliminate competition from foreign workmen, and accordingly the King was petitioned to incorporate a guild for the regulation of the clockmaker's craft in London. In 1631 the Clockmakers' Company was incorporated, and previously to this most of the clockmakers had belonged to the Blacksmiths' Company. The Clockmakers' Company controlled the training of future members of the craft, carefully limiting their numbers so that the market should not be flooded with clocks, but in spite of this London clockmakers were often guilty of having too many apprentices and were accordingly fined. The officers of the Company also had the right to search premises with a constable if it were suspected that watches and clocks of poor quality were to be found there, and if such things were found, the Company had the right to order their destruction. The Company was not one of the wealthier Companies: it never possessed its own hall, and the meetings were usually held in taverns, but during the past three centuries it has numbered among its members some of the cleverest craftsmen in London.

PLATE 4   A German Chamber clock, 16th-century, with foliot
balance iron movement and painted iron dial. Holes in the
wheel carrying the hour hand are for setting the alarm by
insertion of a pin. The hammer hangs over the bell.   *Crown
Copyright. Science Museum, London.*

PLATE 6 (*Above*) **An Automaton clock from Augsburg, early 17th-century. If a receptacle is filled with milk the milkmaid appears to be milking the cow.** (*Left*) **Horizontal Table clock (?) German, mid-16th-century, with Arabic figures.**

PLATE 7 A 16th-century Table clock showing the 'dumbell,' foliot and 'Stackfreed' for modifying the force of the spring without the use of a fusee.

PLATE 8   A Table clock, with astronomical dials; compare
with Plate 1. Nuremburg, c. 1560.

PLATE 9   The opposite side of the clock on facing page
showing the short pendulum – which was added later.

PLATE 10 A little Vertical Table clock by Bartholomew
Newsam, c. 1590; an early example by an English maker. The
little door enables the fusee to be inspected to see if the clock
is near the time to be re-wound. Also note the long tapering
fusees.

PLATE 11 Balance wheel Lantern clock by Thos Knifton at Ye Cross Keys in Lothbury. The feet and finials are plainer than on clocks by Bowyer, and there is a finial above the bell.
Crown Copyright. Science Museum, London.

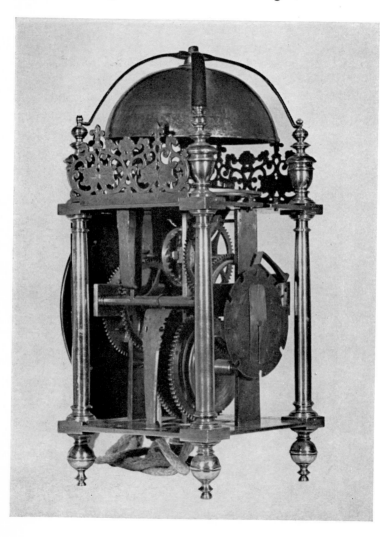

PLATE 12   View of the movement of the clock illustrated on page 39, showing Crown Wheel and Locking Plate.

PLATE 13 A Great Chamber clock by William Bowyer, London. The name appears on the centre of the dial.

PLATE 14 Lantern clock by William Bowyer
(1626–47). Note the bell supported by pin without
decorative finial, balance wheel control, single
hand with tail to assist in setting to time and iron
loops below frame to support weight cords.

PLATE 15   Lantern clock inscribed 'Peter Closon London near Hoburn Bridge Fecit' *c.* 1640.

PLATE 16   Rear view of Plate 14. Note the single spoke to the balance wheel.

PLATE 17 An English Lantern clock by William Bowyer in
Ledenhall Street, 1626–47. Observe the absence of finial above
the bell and wheel balance. The corner finials are very finely
turned.

PLATE 18   An English Lantern clock, inscribed
'Humphrey Clarke of Hartford fecit,' 1670–80.
The 'wings' allowed the anchor-shaped pendulum bob to be visible and indicate that the clock
was still going.

PLATE 19 A Lantern clock with pendulum in
front of the dial, inscribed 'Edward Norris at the
Cross Keys in Bethlem fecit.' This position for
the pendulum is rare in British clocks.

PLATE 20 A late 17th-century travelling strike and alarm Lantern clock. Charles Goode, London. The owner's initials 'C.W.' are incorporated in the dial engraving.

PLATE 21 A pair of late Lantern clocks, showing how the
original style had been modified by the early 18th-century.
(*Left*) a miniature with a silent escapement, by Benjamin Gray,
London; (*Right*) a clock by Richard Rooe, Eperstone.

PLATE 22   This French Lantern strike and alarm
clock of the early 18th-century is less robust than
its English contemporaries.

PLATE 23   A small, early 18th-century travelling alarm of 'Lantern' type, with hoop and spikes at rear for hanging on the wall. Tho. Bucknell, Berkhamsted.

PLATE 24   A Hooded clock, a form in which the Lantern
movement often appeared in the 18th-century. This example is
by John Drew, London, and is probably late 17th-century.

3

THE ADVENT OF THE
PENDULUM

By the middle of the seventeenth century clockmaking had reached a comparatively high standard. Table clocks were being made with various astronomical indicators, and their movements included parts made of brass and were much more finely made than in the previous century. The weight driven wall clock became refined into the Lantern clock in England and in other countries also it was subjected to improvements and new forms of decoration. Many watches were being made and fitted into richly decorated cases of gold or silver or else adorned with rich enamels, and the improved workmanship necessary to produce them was also reflected in contemporary clocks. The only disadvantage was that all instruments of this period were shocking timekeepers. Errors of a quarter of an hour to an hour a day could be expected, and a clock that gained one day might lose the next.

About 1600, an attempt was made to improve on the time-keeping instruments of the day by using the so called 'Cross Beat'. This provided virtually two foliots geared together which moved in opposite directions. Each had its own verge bearing one pallet only, and the scapewheel could now be made flat instead of crown shaped. The arrangement still had no definite period of its own, but the arms of the foliots were slightly flexible and formed an anticipation of the balance spring which was to come in the sixteen-seventies. The Cross Beat was used by Jost Burgi for observatory clocks and to a limited extent by other makers. It was never widely used, and examples encountered so far come from Germany, Denmark, Belgium and Austria. It was probably never used in Britain.

The Italian scientist Galileo is supposed to have noticed that the swinging lamps in the Cathedral at Pisa took the same time to perform each swing whether swinging in a wide or narrow arc. After he had gone blind near the end of his life, he dictated to his son a description of a timekeeper controlled by a pendulum. The son began work on a model of this timekeeper but left it incomplete at his death in 1649.

A German clockmaker named Johann Phillip Treffler was also thinking of a pendulum as a time measurer, and made a clock for the Medici Palace in Florence, but his work is not widely known and it is only the recent researches of Mr S. A. Bedini of the Smithsonian Institution in Washington that have given Treffler the credit he deserves.

The name that will always be associated with the use of the pendulum to control a clock is that of Christiaan Huygens, the Dutch physicist. Huygens made his experimental model on Christmas Day 1656 and obtained a patent from the States General in 1657. He commissioned a clockmaker in The Hague, Salomon Coster, to produce the clocks, and Coster turned out some excellent work incorporating the new principle. The fineness of the workmanship suggests that Coster was used to producing table clocks of very high quality.

The early pendulum clocks broke completely new ground in clock design. These clocks were so far ahead of their time that until recently many people regarded them as Victorian. Most noticeable outwardly was the use of a wooden case, which was almost unknown at the time, while internally there was the pendulum itself and also the use of a direct drive from the mainspring without the use of a fusee. Apparently such confidence was felt in the new controller that it was not considered necessary to modify the force of the mainspring. The movement was virtually that of a table clock turned on its side, while the pendulum was very short, as the verge escapement was retained, involving the pendulum swinging through a wide arc. It was now at long last worthwhile to fit a hand indicating minutes.

The pendulums of these clocks were hung on silken cords which were guided by brass cheeks. After Huygens had made his discovery he went on to discover that the principle of a pendulum taking the same time for long or short swings was not exactly true. The rule was true only when the bob of the pendulum described a curve known as a cycloid, which was slightly more U-shaped than its normal path. The cheeks of the clocks were subsequently amended to interfere with the path of the suspension and make the bob describe a cycloid. One finds clocks of a comparatively late date with cheeks which are not the correct shape for this purpose, indicating that the makers did not know the theory of the idea or else ignored it.

The Huygens-Coster type of clock became very popular in Holland, and the design also arrived in France, where the clocks were known as *Religieuses*. The Dutch name for the type was *Haagse Klokje*. The French examples were basically the same as the Dutch ones but tended to be larger and more highly decorated. The style associated with the Louis XIV period was already beginning to appear.

As soon as the invention became known in England, John Fromanteel, a member of a family of clockmakers of Dutch descent living in London, came to work for Coster to learn how the new sort of clock was made. An invention that brought the accuracy of a clock within a few minutes per day was something that would be eagerly sought after by every clockmaker. By 1658 the making of pendulum clocks was being advertised in the *Commonwealth Mercury* by Ahaseurus Fromanteel, another member of the family, and John Evelyn records a visit to Fromanteel's shop in his diary 'to see some pendules'.

The introduction of the pendulum into clockwork marked the beginning of the period of almost two centuries during which the

London makers led the world in craftsmanship and invention. Not until the mid-nineteenth century, when the London makers refused to move with the times, was that supremacy lost. The restoration of the Monarchy in 1660 meant the end of the period of austerity enforced by the Commonwealth, and the people were ready to spend money to re-furnish their homes in the latest style. Refurnishing the home would in many cases include a new clock; not one of the old wall clocks that were so inaccurate but one of the new ones that kept time so exactly, and of course the case had to match all the new furniture. We find at this period that architectural styles in ebony were the fashion for clock cases, and British makers quickly abandoned two of the main features of the Dutch clocks. They connected the pendulum directly to the verge, doing away with the separate suspension and the cheeks in one blow, and they also did away with the velvet ground to the dial, preferring matted brass or later plain brass engraved and spandrels made of cast brass applied separately.

Development in the late seventeenth century was rapid. The severely architectural styles of the sixties evolved into the basket top of the sixteen-eighties. Ebony remained a favourite wood for cases, but clocks tended to get taller as the century progressed, and movements became more refined technically. No record of this period would be complete without mentioning the name of Thomas Tompion. Tompion forms the subject of a separate book, to which readers are referred for a complete account of his life and work. He acquired the title of The Father of English Watch-making, and was buried in Westminster Abbey. Six thousand watches and five hundred and fifty clocks by him have been listed, which means that he could not have made them all with his own hands. Such production could not be achieved without a large staff of skilled workmen, and it is perhaps as the first production engineer rather than as an horologist that Tompion ought to be remembered. Rumour has it that he established a workshop at Aldgate outside the City limits in order to escape the jurisdiction of the Clockmakers' Company. His official premises and residence

Spring driven pendulum clock

was established at Water Lane, Blackfriars, near Fleet Street.

Neither Tompion nor any of his contemporaries made their own dials or cases, each being provided by skilled engravers or cabinet makers. The name of the 'clockmaker' on the dial indicated the man who supervised the production of the clock and not the man who actually made it. As time progressed, we find specialist workshops producing movements for more than one celebrated maker who would gaily put his name on the dial as if the work had been his. He took the responsibility for the finished article, but was maker in name only.

The spring driven pendulum clock began in Holland but was developed in London, and gradually other European countries produced their own versions of the design. Italy and Austria tended to follow London designs and Bavaria developed a style of its own known as Bamberg which although in general harmony with the English product stood on small feet and incorporated more curves in its design than English clocks were accustomed to possess. Holland tended to develop along British lines although Dutch clocks usually have additional complications such as tune playing or moving figures on the dial.

All these changes, of course, took a long time and lasted well into the next century. France seems to have been little affected by what happened in England and produced a school of clock-making completely different from the English. Germany developed the spring driven clock as a wall clock with a short pendulum swinging in front of the dial which was made of embossed metal in the form of a dish (*Telleruhr*). The chapter ring of the Telleruhr often resembled that used on the later English Lantern clocks, but some of them were made in a Continental style. The ordinary English type of spring driven clock was not greatly developed in Germany. Austria later abandoned the English fashion and tended to copy the French, possibly because of the marriage of Marie Antoinette to Louis XVI of France. Switzerland also tended to follow French styles and eventually settled on what is known as the Neuchatel style, a balloon-shaped clock standing on a bracket.

Bracket clocks

Mention of the bracket reminds one of the usual English name for spring driven pendulum clocks of the seventeenth and eighteenth centuries, i.e. 'Bracket Clocks'. These clocks were not usually placed on brackets, which implies a permanent home, but were rather intended to be carried from room to room and were even provided with a carrying handle for this purpose. They retained the verge escapement until late in the eighteenth century as it is less sensitive to changes in position and is therefore more useful on a clock that has to have its position frequently changed. Many of these clocks had no striking mechanism but were provided with a cord which could be pulled to make the clock repeat hours and quarters; a very useful feature when the clock was stood on a bedside table during the night. The term 'Table Clock' would be far more appropriate for these clocks, but as it has already been used for the earlier metal cased clocks, 'Bracket Clock' will have to stand.

The advent of the pendulum not only led to a development of the spring clock as first created by Coster, but also produced an entirely new type of clock. The old clocks with foliot or wheel balance would have needed winding every twelve to fifteen hours and at the same time it would be necessary to regulate them every day after comparison with a sundial. Most public clocks had a sundial nearby for this purpose. After the clock had been made so accurate that the daily regulation was no longer necessary, there was an incentive to prolong the intervals between winding. Spring driven pendulum clocks in their early days were made to run for one day, then two or more and finally eight days between windings. During the early years of the reign of Charles II, London makers evolved a new type of clock which would run for eight days. The movement was generally similar to that of the spring clock except that the drive was by weights supported by catgut lines which were wound round brass barrels, and the clock was wound by means of a key through the dial as were the spring clocks. The short pendulum was retained as the verge was still the only available escapement. The ebony architectural case was provided but the clock was intended to be hung on the wall. The Dutch spring clocks were usually provided with means for hanging them on the wall as well as feet for standing on a table, but spring clocks in England were seldom seen with this feature. The exposed weights of the new type of clock were provided with polished brass containers, but even so they were considered unsightly and clocks were produced with a long tall cupboard below them to hide the weights from view. The next step was to make the cupboard and the top into a free standing unit and the 'Long Case' clock was born.

The earliest were quite small being only about five feet high, and the ebony architectural style was used for the cases. The Fromanteel family were associated with this type in the early days, and may even have been the pioneers of the type. The movements were closely allied to the pendulum controlled spring clocks which were being produced and which were direct descendants of the Table clocks. The Long Case Clock is therefore more closely connected with the Table clock than with the weight driven wall clock.

The advantages of a longer pendulum were being considered at this period. A longer pendulum would be more capable of receiving fine adjustments in length and therefore the regulation of the clock could be more exact, as one turn of the rating nut below the bob would make far less difference in proportion than on a short pendulum. By having fewer beats per hour fewer teeth would be necessary in the wheels, thereby saving labour. Robert Hooke, the Curator of Experiments of the Royal Society demonstrated in 1669 that a very long pendulum could be kept swinging by the impulse given by a pin in the rim of the balance of a pocket watch, although the arc was very small. This was of itself an advantage, for when a pendulum swings through a very small arc the difference between the circle and the cycloid is small enough to cause no concern.

The verge escapement required a very large arc, say 40°, and could not have been used with a long pendulum swinging that far.

Anchor escapement

The problem had been tackled in Holland as will be related in the next chapter, but the solution would not have been acceptable in the construction of a Long Case clock. The only way to produce a satisfactory answer was to provide an escapement that only needed a very small arc to unlock. Mr R. A. Lee has discovered a clock by Knibb which employed a variation of the Cross Beat mentioned previously, used in conjunction with a long pendulum. This had the disadvantage of involving pivoted levers which not only increased the complexity of the mechanism but which were theoretically unsound. The true answer was provided by the invention of the anchor escapement.

There has been a lot of discussion on the subject of whether Hooke, mentioned above, or William Clement, a London clock-maker, was the inventor of the anchor escapement. Clement seems to be winning at the moment. A very strong point in his favour is that he was originally an anchor smith, and the shape of the anchors on which he worked no doubt suggested the escapement. As well as making clocks with a pendulum beating seconds, Clement produced some with a pendulum five feet long beating a second and a quarter. These early clocks had cases very little larger than the short pendulum type, but as the century progressed there was a tendency for the Long Case clock to get bigger, especially when examples were produced which ran for one, three or six months between windings.

The earliest clocks with a very long period of running were three made by Tompion in 1676 for Greenwich Observatory. These clocks were intended to be wound only once a year. They were mounted in a space inside the wall, and therefore hardly come in the Long Case category, but they led the way to the clocks made about the end of the seventeenth century with three, six or twelve month trains which were very large indeed compared with the clocks of thirty years previously.

The Long Case clock rapidly became popular after the invention of the anchor escapement. The cases at first had an ebony finish and later marquetry and parquetry became popular, while in the eighteenth century lacquered cases were fashionable. In the early part of the same century walnut also occurs, and once mahogany had established itself its popularity lasted until the end of the English Long Case clock in the nineteenth century. The Lantern clock was still being made in the eighteenth century, but many movements of the Lantern type were also being produced with square brass dials intended to be covered by a hood and hung on the wall, or else to be fitted in a country version of the Long Case. One encounters many of these one-handed clocks in a variety of cases not always pleasing to the eye, and in many cases a minute hand has been added. This can always be detected when the old dial is retained, for the original dial has quarter hour marks inside the hour figures and no minute marks outside them. Some of these clocks have movements with the wheels held between two brass plates as on the eight-day clocks, but they are still to be considered as belonging to the Lantern family. In their final form down to about 1850 they were produced in Birmingham with the usual painted iron dials found on nineteenth century clocks. By this time most of the eight-day movements and dials were being produced there, with the name of the vendor painted on the dial to order, and the cases would be made to the order of the purchaser in his own locality..

As the Long Case clock increased in size, the dial tended to increase also. Beginning with a size of about nine inches square immediately after the Restoration, dials had increased to twelve inches square by about 1690, and early in the eighteenth century an arch was placed above the square dial. Early arches were made separately, but soon the dial and arch were being made in one piece. One of the earliest arch dials is to be found on the Tompion clock in the Pump Room at Bath. Dials at this period were made of sheet brass with silvered chapter rings and cast spandrels fitted separately, but as the eighteenth century progressed some dials were made with the figures engraved directly on to the dial itself, or else the dial was silvered all over and the figures were indicated in black. This led the way to the painted iron dials characteristic of the late eighteenth and early nineteenth centuries.

As soon as the arch was added to the dial it became a space that needed filling. Sometimes a plaque with the maker's name and place of business was put there, or a strike/silent hand would occupy that position. After about 1730 the phase of the moon was indicated by a disc bearing two moons that rotated once in two lunations, and this neatly filled the space in the arch. The square dial was mostly used on the thirty-hour Lantern type movements, but had a longer life on eight-day clocks in Lancashire and Wales where it would sometimes include moonphase in an opening below the figure XII. To know the phase of the moon was of great importance in the eighteenth century, as only at the time of full moon was it possible to go out at night. Moon dials often include an indicator to show high tide at a certain port, but this has to be specially calibrated for the place in question and is of no use elsewhere. The day of the month was indicated at first by a figure showing through a square opening above figure VI. Later a dial was placed here or a kidney shaped slot allowed figures on a disc to be visible. The seconds hand placed below figure XII goes back to the early days of the anchor escapement, for with a pendulum beating seconds and a scapewheel of thirty teeth the division of the minute can be indicated on the dial with very little extra trouble. The eight-day clock is recognisable at a glance by the winding holes in the dial, but even this can prove a trap for the unwary. Less fortunate people who could not afford an eight-day clock would sometimes have false winding holes painted on the dial of their clock to give the impression that it was an eight-day one when in fact it was only a thirty-hour clock and would be wound by pulling the chain or rope inside the trunk.

A Long Case clock can be closely dated by observing various features of case, dial and movement, but there is not space in a work of this type to treat the subject in detail. Readers desiring further information are advised to consult the Bibliography at the end.

The Long Case clock became particularly associated with

England. The only other country that took it seriously was Holland, and tradition was still strong enough to keep it popular in the U.S.A. after 1776, although native American styles drove it off the market there before it had disappeared in England. In the latter country, its popularity in London waned during the later part of the eighteenth century, but it was still popular in the provinces, particularly in the north, and the mill owners made rich by the Industrial Revolution always desired very flamboyant styles for their homes, helping to create distinctive types for Lancashire and Yorkshire as the eighteenth century gave way to the nineteenth. The type was also popular in Wales and Scotland.

As far as the other countries in Europe were concerned, the Long Case clock did not achieve the popularity it enjoyed in Britain. Many countries however, produced their own version of the design but always in limited quantities. Holland showed more interest in it than most, and Belgium produced some Lantern type movements in long cases. These were made in the Flemish and French speaking parts of the country and in the latter part are known as 'Liége'. Sweden and Denmark produced clocks in the style of the British ones comparatively late in the eighteenth century, while Austria put lantern type movements in long cases with the short pendulum swinging before the dial as on the Telleruhr. In Austria also, plain versions of the eight-day Long Case clock were produced in the early nineteenth century. Austria eventually produced a refinement of the Long Case clock in the Vienna Regulator. These clocks always hung on the wall, generally had pendulums beating more rapidly than once per second, and the fineness of their movements approached the standard of watchwork. Judging by the tempo of the second movement of Haydn's *Clock Symphony*, the composer probably had an early example of a Vienna Regulator in mind when the Symphony was written. The Austrian Emperor Francis Joseph I (1848–1916) had a Vienna Regulator in his study at his country residence.

The early Vienna Regulators had a pronounced hood, trunk and base, and in shape resembled a Long Case clock. Their cases were glazed on three sides, giving a view of the pendulum and weights. Later clocks of this type had a straight sided case, and the affinity with the Long Case clock was less marked. Various sizes were produced, the smallest being about eighteen inches high and the largest almost as big as a normal Long Case clock.

The type first appeared in England at the Exhibition of 1862. Lord Grimthorpe, one of the foremost authorities on horology of his time, and designer of Big Ben, condemned them on the grounds that their pendulums moved through a very small arc and it made them prone to stop on very slight interference. It has been known for a spider's web spun between the pendulum and case to cause a pronounced acceleration on one of these clocks.

North Germany produced some Long Case clocks in the late eighteenth and early nineteenth centuries, but although they generally followed the English plan, their cases had the hood, trunk and base of the same width, and contained more decoration in the form of carving. The dials were usually of the arch form with an embossed metal ground having an enamelled circular portion in the centre for the figures.

About the same period Sweden had evolved a type which is usually known as the 'Farmhouse' clock and had quite a lot in common with the English style of Long Case clock, as far as the movement was concerned, while the case derived much of its inspiration from France. These clocks had a subsequent influence on American designs.

France produced Long Case clocks, but not in the English tradition. Early examples appeared to consist of a mantel clock standing on a pedestal, which in fact was the lower part of the case and housed the pendulum. In the early eighteenth century the so-called Comtoise made its appearance. These clocks had an eight-day weight-driven movement with the wheels arranged between vertical bars as on the early weight-driven wall clocks. The trains were arranged side by side, but with the going train on the nine o'clock side instead of on the three o'clock side as in English clocks. Often the clock would be arranged to repeat the hour strike two minutes after the hour. The pendulum often hung at the front instead of at the back of the movement, and the verge escapement was used with the scapewheel upside down, and very long pallets were fitted to keep down the amplitude of the pendulum's swing. The movement was cased in a sheet iron box, and could be used as a wall clock or fitted with a standing case as desired. In the latter instance the door of the case was often glazed and the pendulum decorated in an extremely elaborate fashion. The basic type of pendulum was jointed to allow for easier travelling.

The Comtoise was a country clock made in the district of Franche Comté, and it enjoyed a very long life. It is believed to have been in production as late as 1914, but the later examples had anchor escapements. Some of these clocks are known which possess quarter chimes. For the more important French towns, a more sophisticated version of Long Case clock was evolved with a high quality movement, and the cases were often decorated with Boulle or Ormolu. The trunk was narrow at the top but swelled towards the bottom where the pendulum bob swung. A glazed door was also usual with this type. The dials were generally smaller than on English clocks and circular, being enamelled. Enamelled dials were popular on French clocks. Originally each figure was contained on a separate plaque, but as technique improved, the whole dial was made in one piece.

The Long Case clock is often known as a 'Grandfather' clock but the term is comparatively modern. It originated about 1880 as a result of a popular song by Henry C. Work which went 'My Grandfather's clock was too big for the shelf, so it stood ninety years on the floor.' The name seems to have come to stay.

PLATE 25 A clock with a cross beat escapement *c.* 1630. Two arms with cherub's head terminations can be seen in the upper part. Indicators are provided for moonphase, calendar and zodiac.

PLATE 27 A very early English spring-driven pendulum clock. Note the high position of the winding holes, and the dial left plain except for the Tudor Rose in centre.

PLATE 28   An early Bracket clock movement by Edward East,
London, 1660–70. It has a very small locking plate, and
'balustrade' pillars.

PLATE 29 An architectural Bracket clock by Edward East, London, 1660–70. Note the high position of the winding holes and the day of the month aperture.

PLATE 30   A Bracket clock by James Markwick, London,
signed 'Jacobus,' c. 1670. Time strike alarm and day of month.
The hole between figures I and II is for the alarm winding
square. Note cherub's head spandrels.

PLATE 31   A Bracket clock by John Small-
wood, Lichfield, 1756. Has an ebonised case
and an arch dial.

PLATE 32   An English Bracket clock by de
Charme, late 17th-century. Note the square dial
and the cherubs head spandrels.

PLATE 33   A large three train Musical clock, 'Eduard
Pashler' *c.* 1775. It has a bell top mahogany case and plays
six alternative tunes.

PLATE 34   A Regency Bracket clock of the early 19th-century. The dial has become circular, the dummy pendulum has disappeared, but the carrying handle and cord for pull repeat are still present.

PLATE 35 A Regency Bracket clock by French, Royal Exchange London c. 1820.

PLATE 36 This small timepiece by Arnold, 84 Strand,
c. 1830, has a rosewood case with carrying handle, and
an enamel dial with gilt surround.

PLATE 37   Side view of Long Case movement, *c.* 1670.   Observe the thickness of the spandrels.

PLATE 38   An early Long Case movement with a short pendulum, Joseph Knibb, c. 1670. Note the large size of the wheels attached to the barrels, the latches for fastening dial to movement, and the hammer inside the bell moving horizontally to strike.

PLATE 39 Model of the anchor escapement used in English Long Case clocks. *Crown Copyright.* *Science Museum, London.*

PLATE 40   The movement of a late 17th-century English Long
Case clock, Jas. Clowes, London. Note the locking plate which
rotates with the barrel and the very small scape wheel.

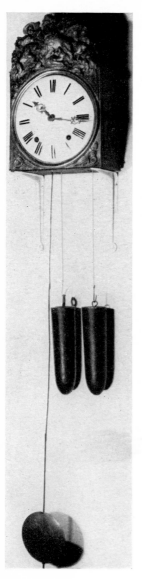

PLATE 41 (*Above*) **A German 'Telleruhr' developed as a standing clock. Early 18th-century.** (*Left*) **A 'Comtoise' arranged as a wall clock. Note pendulum at front of the clock with joints in the rod to facilitate transport.**

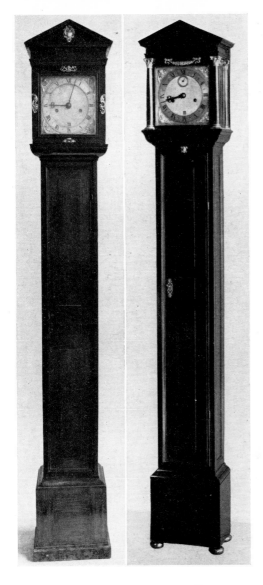

PLATE 42 (*Left*) An early 'architectural' style Long Case clock by Ahaseurus Fromarteel *c.* 1670. The day of the month is indicated above Fig. VI. (*Right*) An early example of a Long Case clock with seconds dial by John Fromanteel, *c.* 1675. The day of the month is indicated above Fig. VI.

82

PLATE 43 (*Far Left*) **Month Long Case clock by David Guepin, London. With marquetry case and bold design of the hour hand,** *c.* 1690.
(*Centre*) **Quarter chiming Long Case clock by Henry Jones, London,** *c.* 1690. **Has marquetry case. Note the convex moulding below the hood and window in door to allow the pendulum bob to be seen.**
(*Right*) **Long Case clock by George Graham, London, 1730–40. Has an 8-day striking and equation movement. The case is of oak veneered with walnut. Note the unusual position of the seconds dial.**

PLATE 44 (*Left*) **A mahogany Long Case clock, made by Joseph Stevens,** *c.* 1780.
(*Right*) **A Long Case clock with a painted dial. Chune of Shiffnall,** *c.* 1820

PLATE 45   A 'Telleruhr' by Marcus Böhm, Augsburg, c. 1700.
Compare the pendulum bob with the ends of the cross beat
arms on Plate 25.

4

CLOCKMAKING IN HOLLAND

The origin of the Long Case clock is claimed for London, but it is more likely that it developed in Holland at the same time. Seventeenth-century Dutch Long Case clocks show a great similarity to those of England, the main difference being the retention of the velvet ground to the dial that appears on the Coster type clock. By the time that the eighteenth century styles had become established. Dutch makers were especially fond of making their clocks as complicated as possible. They introduced musical work with a subsidiary dial to vary the tunes, moonphase, time of high water, day of the week with a representation of the appropriate deity etc., and these complications not only affected the Long Case clock but also the Bracket clock. On both types we also find moving figures which either performed continuously in time with the pendulum or only when the striking or musical work was in action. A rocking ship would appeal particularly to the Dutch, and other features found on Dutch clocks were the windmill with rotating sails, and the fisherman who kept catching the same fish over and over again. The little scenes where these figures moved were decorated with paint, and marked a break away from the exclusive metal finish that was applied to the dials of English clocks. English clockmakers used paint for their moon discs, but it was frowned upon generally until quite late in the eighteenth century.

The cases of Dutch Long Case clocks are usually more elaborate than their English counterparts, often having bases of a bulbous shape with elaborate paw feet, and hoods surmounted with carved figures of Atlas holding the world flanked by angels with trumpets. Marquetry continued well into the eighteenth century after London makers had abandoned it, and it occurs in Holland in conjunction with arched dials, whereas in its London period square dials were the rule.

Towards the end of the eighteenth century, French styles began to be popular in Holland, and in the large towns the making of clocks of the English type declined. To find really distinctive Dutch styles we have to go to the country and pick up the narrative at the time of the invention of the pendulum in 1657. The first clocks with pendulums were the small finely made spring driven clocks produced by Coster under Huygens' patent. In 1658 Huygens and Coster were experimenting with pendulum control of turret clocks, and the clock on which they conducted their experiments was the church clock at the fishing village of Scheveningen near The Hague. The clock itself is now lost, but the pendulum and escapement have been preserved at Huygens' country residence, 'Hofwijk', a little to the east of The Hague, and can still be seen in action there.

The making of turret clocks in Holland had reached a very high standard, and the frames were generally elaborately decorated. A magnificent specimen from the St Jacobskerk at The Hague has been preserved in the clock museum at Utrecht, and includes the carillon which was such a feature of the Low Countries. The elaboration of this movement is greatly in advance of English work of the period. It is of course, controlled by a verge escapement and foliot which means that its performance would have left much to be desired. The Dutch have always been conscious of the need for public timekeepers, and the visitor to Holland today will be impressed by the number of towers to be seen which possess clocks. Once the pendulum had established itself, the Dutch people would be anxious to bring the benefit of the

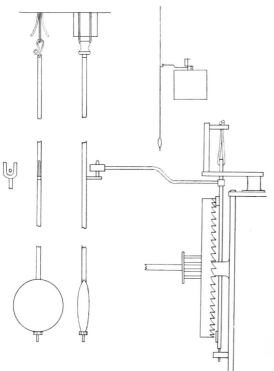

A Turret clock movement with Verge Escapement converted to pendulum control by the method used by Huygens at Scheveningen. The small upper sketch shows the general arrangement.

new invention to public clocks, hence the experiments of Huygens and Coster.

The method they used to apply the pendulum to an existing turret clock was to suspend it near the movement with its point of suspension placed so high that it hung with its middle point roughly level with the top of the verge. The foliot was removed and replaced by a long rod which engaged the pendulum rod near this middle point. By this means, the amplitude of the swing was kept small, and provided that the theoretical number of beats to the minute of the original foliot was not too small, a long enough pendulum could usually be arranged in the tower. If the beat were slower than that of the pendulum, a scapewheel having more teeth would have to be made. To give an example of this, the Salisbury clock beats theoretically once in four seconds. If it were converted by the Huygens method to pendulum control, it would need a pendulum 52 ft long supported 26 ft above the movement. The easiest solution in this case would be to add another wheel to the going train and provide a shorter pendulum, but this would cost more money.

The Huygens method of conversion proved successful, and not only were old clocks converted but new ones were made on the same principle, and eventually the method found its way into domestic clocks. For the development of the latter we must turn our attention to the district north of Amsterdam drained by the river Zaan, with its centre at Zaandam. This area was industrialised by the seventeenth century, specialising in shipbuilding, timber sawing, processing of various agricultural products, etc., and a class of wealthy industrialists arose who demanded a high quality clock. There was a reluctance to abandon the weight driven wall clock, and a new type appeared in this district that is consequently known as the *Zaanse* clock. Weight driven wall clocks at this time often had their movements cased in sheet iron with designs painted on it, but the new clock had a wooden case with a high quality finish, and was arranged to look as if standing on a bracket which extended well above and well below the clock. The bracket was in fact hollow, and contained the pendulum which had its halfway point about level with the top of the movement and was arranged as on the converted turret clocks.

The wheels were of brass and the sides of the case were glazed so that they could be seen. The dial with its velvet backing and silver chapter ring suggested the Coster pendulum clocks, and sometimes a separate dial indicating the quarter hours would be fitted below the main dial. The bell would be placed over the clock and have a statuette mounted on it – usually Atlas bearing the world on his shoulders or alternatively Minerva

— and would be flanked by cast brass frets. The pendulum bob would be visible in the opening below the bracket, and usually consisted of some such subject as a man on a horse. The weights were suspended on cords and were made pear-shaped with polished brass cases.

These clocks were made from about 1670 to 1720 and are now highly prized collector's items. They often incorporate double striking, i.e. the full number of the following hour is sounded at half past on a smaller bell. This feature is popular on Dutch clocks generally, as the Dutch method of reckoning time is to say that half past one is 'Half two', half past two is 'Half three' and so on. Dutch striking has also been included on some English seventeenth century clocks, but English clocks generally sounded full hours only.

One occasionally meets faked Zaanse clocks which were prob-probably produced in the late nineteenth century using parts for Frisian clocks (see below). It is often difficult to make up one's mind about a given example, but the thinness of the wheels usually gives them away. The wheels of a genuine example are extremely thick. These clocks should not be confused with the modern reproductions which do not set out to be anything but reproductions and have modern style movements.

The next area of Holland to be considered is Friesland. This province was, and is, renowned for Dairy Farming, but a definite school of clockmaking established itself there. The type of clock particularly associated with Friesland is the *Stoelklok* (Chair clock) and this can be considered as a simpler version of the Zaanse clock. The people of Friesland maintain that the Stoelklok is older than the Zaanse clock, being originally made about 1600, and that the Zaanse clock was derived from the Stoelklok, but the matter will probably never finally be decided.

The Stoelklok was supported by a bracket hung on the wall and covered by a small roof to keep the dust from the movement. The roof itself was sometimes covered by a lace or linen runner the ends of which hung down on each side of the clock (*Klokkek-*

Chair clock

leedje). The clock itself stood on a small platform which was supported by the bracket, the platform having turned feet. The bracket and dial would be brightly painted, and brackets tended to fall into two groups as far as design was concerned. These were the 'Vase' type where the sides of the bracket were ornamental cut-out shapes like the handles of a vase, but which also suggested the double eagle of the Holy Roman Empire, and the 'Mermaid' type which had a brightly coloured mermaid on each side. The dial of the clock was also brightly painted and both bracket and dial were flanked with cast lead frets which were gilded and sometimes painted as well.

The movement was basically similar to that of the Zaanse clock except that the top and bottom were formed of sheet iron painted with red lead, and that chains with links like a figure 8 were used in place of the cords. The pendulum was shorter and swung in the space between the back of the movement and the bracket, the ticking of the clock being accompanied by a rhythmic scraping noise caused by the pendulum and the wire connected to the verge rubbing together as they pursued their widely differing paths.

The striking usually provided for hours in full with one blow at the half hour, but examples with double striking are known. Alarm work was generally fitted, but complications such as moonphase or day of the month were rare. When the alarm was not required to sound, the alarm weight and its counterpoise would be looped up, making a very attractive decoration.

The Stoelklok is generally believed to have appeared about 1700, i.e. after the Zaanse clock was well established, and the earliest examples had an hour hand only. Dials have often been repainted, and one may find a dial with an hour hand only where the minutes have been added. It was possible in the nineteenth century to buy paper dials to stick on these clocks to save the expense of a repaint, and these dials were provided with minute marks and also had quarter hour marks inside the figures to suit single handed clocks.

Tail clock

The Stoelklok was generally associated with Friesland but other provinces of the Netherlands had their own versions. Drenthe was fond of the figures of Faith, Hope and Charity for the frets, Groningen had more elaborate carving and generally more solid movements, North Brabant replaced a lot of the turned brass decoration by plain iron parts and so on. A feature that is sometimes met with in Drenthe clocks with a single hand is that the clock is arranged not to strike at 12.30 and 1.30, a single blow between 12 o'clock and 2 o'clock being reserved to indicate 1 o'clock.

The Stoelklok was made in large quantities, and various specialist crafts contributed towards its making. There were brass and lead founders, cabinet makers, dial and bracket painters etc., while the clockmaker as such, working from prepared material, cut the teeth of the wheels, mounted them in the frame and produced the finished clock.

At the beginning of the nineteenth century the Stoelklok tended to become simpler and many were made with time and alarm only. The workmanship was rougher than had previously been the case. As the century wore on the large scale production of these clocks ceased, although they could still be obtained to order. Even today, specialist firms can produce these clocks in the traditional style.

One reason for the decline of the Stoelklok was the more sophisticated *Staartklok* (Tail clock) which appeared about 1800. The basic movement of the Stoelklok was retained, but the verge escapement was replaced by an anchor escapement with a long pendulum. The movement was covered by a hood as in a Long Case clock, and provided with an arch dial. Although the pendulum was enclosed in an elaborate box reminiscent of the Zaanse clock, the weights hung in the open. The pendulum bob showed through a window which was decorated with a pierced brass plaque. The Staartklok was intended to be a cheaper version of the Long Case clock, and although most examples are plain, one finds others indicating moonphase, day of the month, and possessing mechanical moving figures such as their more illustrious counterparts. Although the idea was to imitate the Long Case clock, the Staartklok still hung on the wall, and was a relative of a type known as the *Amsterdammer*, which tried to combine the features of a Long Case clock in a wall clock, still retaining a movement of Long Case quality. These clocks are very rare.

The Staartklok was generally made in four sizes. The very largest usually possessed some extra feature such as double striking or moonphase, and often mechanical figures or music would be incorporated. The second size was the type that is most frequently met with and generally limits its special features to day of the month and moonphase. The third size was like the second in shape but scaled down to a length of about 3 ft 3 in. This size is known as the *Kantoorklok* (Office clock), and being rare is keenly sought after by collectors. The smallest size of all dispenses with the long pendulum inside the box and has a short pendulum inside the hood together with a verge escapement. As these clocks were intended for use on vessels on Dutch inland waterways, this type of escapement lent itself better to the movement of the vessel. Sometimes the pendulum is arranged to swing fore and aft instead of from side to side. The clocks are known as *Schippertjes* (little ships' clocks) and are highly desirable from a Dutch collector's point of view. The miniature Stoelklok that one occasionally sees was probably intended for the same purpose.

The Staartklok had a fairly long life, being made roughly from 1800 to 1880. It seems to be a Frisian speciality, for the other provinces obtained their clocks from Friesland. They were even exported from Friesland. A short trunk version of the standard model was nicknamed 'Turk' as it formed an export commodity to Mediterranean countries.

Clockmaking in Friesland tended to concentrate itself in later years in the town of Joure, and this town is particularly associated with the Staartklok, only a few being made elsewhere. A variety with the normal type of hood but no *Staart* or tail is known as a 'Mechelse' clock, taking its name from the town of Mechelen (Malines) in Belgium. Peak year for production in Joure was 1857, when four thousand clocks were produced, but by 1880 or so the Joure industry was all but dead, killed by imports from the Black Forest factories which had begun production on American methods. Fashion played a part too. People regarded a factory made clock as superior to a hand made country production and at the end of the life of the Dutch product skilled men were being paid the equivalent of a few pence per hour. Not only in Holland were the old craftsmen receiving low wages, but even in the Black Forest itself the makers of the traditional type of clock produced there were being forced down to lower prices and lower wages and were eventually obliged to give up business altogether because of the all-conquering factory.

About this time the Dutch tradesmen were stimulating sales of German clocks by allowing a rebate on old Dutch clocks brought to them when a sale was made. The old movements were then sold as scrap for the melting pot, and the cases and brackets chopped up for firewood. One consignment of these old Dutch clocks actually reached England where it lay forgotten in a warehouse in Croydon for about forty years, and was then discovered by a clockmaker who purchased it and subsequently put the clocks into circulation again. Several of them even returned to Holland where they were restored and are now as good as new.

Along with making traditional types of clock to order, a number of firms in Holland are making reproductions of the old types of Dutch clocks, using German movements of the type fitted to cuckoo clocks. These reproductions always give themselves away by displaying two weights instead of one. The smaller varieties are popular such as the Schippertje and Kantoorklok, although a few larger ones are made. Zaanse clocks of all sizes and qualities are made, but are usually smaller than their prototypes. The latest addition to the range of models is the Amsterdammer. These reproductions are by no means cheap, but genuine old clocks are dearer still. Holland is very antique conscious and old clocks fetch high prices. There is a flourishing trade in Holland for importing old clocks from England and selling them to collectors.

Holland is probably unique among the countries of Europe in still offering old types of clock made to traditional designs and by traditional methods. The clocks which are produced to order cannot really be called reproductions, for they are produced in small workshops under much the same conditions and sometimes even with the same tools as in the eighteenth century. The only differences that have been incorporated are an improvement in tooth profile, for some of the old wheels did not run at all smoothly, and the use of silver steel instead of wrought iron for pinions, as it possesses better wearing qualities. The clocks which are being made to the old designs at the present time will certainly last as long as if not longer than their prototypes.

PLATE 46   The movement of a Dutch wall clock with balance
wheel, *c.* 1653. The arrangement of the alarm mechanism above
the going train is similar to that used in later Friese clocks.

PLATE 47 A Haagse Klokje by Salomon Coster, 1657. Every minute is numbered. The dial is covered with blue velvet.

PLATE 48 A *Religieuse*, 'F. Gilbert, Angers', c. 1675. It has black velvet on the dial. Note the similarity to the Dutch style.

PLATE 49 'Haagse Klokje' by Jan Van Ceulen, The Hague,
late 17th-century. Red velvet ground to dial. Bas relief of
Time with arms raised, Architectural pediment. The end of the
winding square is decorated. *Crown copyright. Science Museum,*
London.

PLATE 50   A Bracket clock by Roger Dunster, Amsterdam,
c. 1740. British style, with matted dial centre, dummy pendulum
and silvered chapter ring.

PLATE 52 (*Above*) **An 18th-century, single-striking 'Schippertje' with alarm. In 'Stoelklok' form.**
(*Left*) **A typical Dutch 18th-century Long Case clock showing Atlas and angels on hood, day of the week and month indicators, moonphase, etc.**

PLATE 53 A later Stoelklok fitted with an alarm but no striking work. This shows the more austere models produced in the early 19th-century, when the 'Stoelklok' was being superseded by the 'Staartklok'. *Crown Copyright. Science Museum, London.*

PLATE 54 (*Right*) **This Dutch Staartklok, *c.* 1850, strikes hours and half hours and is fitted with an alarm.** (*Left*) **A large Staartklok with double cap, double striking including one blow at the quarters and alarm. The day of the month is indicated by a cencentric hand.**

PLATE 55 (*Left*) The movement of a Dutch 'Staartklok'. The sheet iron plates at the top and bottom of the movement are painted with red lead. The large scapewheel is prominent, also the rack and snail striking, which here has been put at the back of the clock instead of its usual place just behind the dial. The toothed disc at the back of the dial is for indicating the phases of the moon.

PLATE 56 (*Right*) An early 19th-century 'Schippertje', Time and strike only. It has a painted dial and miniature figures of Atlas and trumpeting angels above in imitation of 18th-century Dutch Long Case clocks.

PLATE 57   Three types of Staartklok awaiting repair. The one
on the left has Father Time as the motif of the pendulum
window while that on the right is the more usual Urn pattern. The centre clock is a 'Kortkast.'

PLATE 58 (*Left*) **A large Stoelklok with mermaid type bracket** (*Right*) **A large Stoelklok with vase type bracket but mermaid decorations to dial.**

PLATE 59 A 'Kortkast' (short case) 'Staartklok' (nicknamed 'Turk'), 1800–25. Double striking with one blow at the quarters. Has a moonphase and date indicator and an oak case.

A 'Zaanse' clock, Groot, 1725. This clock is double-striking with one blow at quarters.

5

CLOCKMAKING IN THE BLACK FOREST

The Black Forest area of Germany lies in the two provinces of Baden and Württemburg. It suffers from very cold winters with heavy snowfalls which render farming impossible, and before the advent of modern transport it was very inaccessible. All the more wonder that it should have developed an export trade in clocks that reached as far afield as Russia and the U.S.A.

The bleak winters made it necessary for the farming population to have some second occupation which could be followed when farm work could not be undertaken. The area had a reputation for glassware, and tradition has it that a travelling salesman for the glassware brought home a wooden clock from Bohemia some-where about 1640. It was a very simple affair, three wheels in the train, verge escapement and foliot, and no elaborations, but the idea was immediately seized upon as a further occupation for the winter months. Tradition also has it that the earliest clocks in the Black Forest had the teeth of their wheels carved with a knife, but this does not seem feasible after one has tried to do it oneself. In some way, the peasants learned the craft and evolved the neces-sary tools, and then production steadily increased till by the end of the eighteenth century a regular export trade had been built up.

The Black Forest clocks were made by a system of divided labour. Each worker would specialise on one particular part; for instance the frame maker would build the frames and send them to the clockmaker proper who would mount the wheelwork and other parts into the frames and fit them with dials obtained from the dialpainter, who in his turn would purchase dials from the dialmakers. The dialmakers would obtain the circular portion bearing the figures from the dialturners. Subsidiary trades would include brass founders for casting the wheels, wheel turners for finishing them, chain makers and gongmakers. Supporting all the other trades would be the toolmakers. All clockmakers had their own standards. There was no question of one man's parts fitting another man's clock.

The earliest Black Forest clocks with foliot are now being reproduced commercially as a tourist attraction. They generally bear the date 1640, but differ in detail from their prototypes, usually having solid pinions instead of the lantern type. Several variations of the original model have been evolved, including one with a 5 ft pendulum which of course has no precedent from the seventeenth century. At the Columbian Exposition of 1893 there were sold small foliot clocks resembling the Black Forest type. These were inscribed with the date 1492 but were an American factory production.

The Black Forest abandoned the foliot very late, possibly about 1740. It was not until 1720 that striking work was incorporated in Black Forest clocks, and wooden wheels lasted until about the end of the eighteenth century as local casting of brass did not begin until 1780 in Neustadt. Previously made clocks with brass wheels would have had them cast elsewhere. Weights were always suspended on cords until 1790 when they were gradually replaced by chains. The sprockets over which the cords passed were cut with teeth in a V-shaped channel to give the cord something to grip. Some clocks made quite late which have chains for the going and striking work (where the latter is fitted), still retain cord drive for the alarm. The verge escapement was still being made as late as 1820.

By the early nineteenth century the Black Forest wall clock

was being sold all over Europe and had also penetrated to Russia and the U.S.A. The clocks were carried around by travelling salesmen who hawked their wares from a *Tragstuhl* or carrying frame. These salesmen would hire a room in a district where they wished to sell, have the stock moved to this room and then go out every day with the Tragstuhl full until all the clocks were sold. The men would be away from their homes for many months at a time, and would be joyously greeted by their families on their return. These vendors of clocks have often formed the subject of a clock themselves, being made in the form of a statuette about 18 in. high with one of the clocks carried by the man containing a small spring-driven movement which acts as the actual clock.

The Black Forest movement, after striking work has been added, consisted basically of a top and bottom of beechwood, with four corner uprights, also of wood, and three vertical wooden plates with brass bushes to carry the pivots of the wheels. In its earliest form the clock would have had wooden wheels, and as the number of teeth that can be cut on a wooden wheel is limited, each train would have had five arbors. When brass wheels came in this would have been reduced to four, but the arbors themselves remained wood with wires at the ends to form pivots. The wooden arbors were usually painted with metallic paint to try and persuade the customers that they were metal.

When the pendulum was first introduced into the Black Forest it was very short and used in conjunction with a verge escapement. It swung in front of the dial, and was nicknamed the *Kuhschwanz* (Cow tail). The anchor escapement and long pendulum did not appear until about 1750. Dials were generally of painted wood, but during the eighteenth century it became fashionable to use paper dials which were hand coloured and glazed over. The side doors of the movement were of the local fir wood and were rather brittle, so they are often missing today. It is not generally realised that the wood for the frames had to be brought from elsewhere as the local wood in the Black Forest is nearly all coniferous. All the iron used in the movements, usually in the form of wire, had to be brought from outside as well as the wood for frames.

The Black Forest wall clock in its final form was a very reliable timekeeper. Hasluck in the *Clock Jobber's Handbook* of 1887 calls the clock 'a most trustworthy timekeeper' and goes on to say that 'no other country makes clocks of this type'. Some were sold at fifteen pence retail, so the prime cost must have amounted to only a few pence.

An important event in the history of the Black Forest clock came about 1730 when Anton Ketterer of Schönwald invented the 'cuckoo' clock. In its early years, the cuckoo clock resembled all other types of Black Forest wall clock, and it is only since about 1870 that it has taken on the form in which we know it today, and which is sold so widely in Europe. Ketterer's invention was quite simple, being based on the church organ pipe, and he probably little realised at the time how popular the cuckoo clock was going to be in the years to come.

The shape of the Black Forest dial consisted for many years of a rectangle with a semicircle above it; the same shape that was used for dials of English Long Case clocks. The painting of the dial usually took the form of bunches of flowers, but the form of the decoration was modified to suit the eventual customers. For instance, the French market liked large bunches of flowers in bright colours and the clocks were sold under the name of 'Swiss clocks'. Belgium and Holland liked tin or porcelain dials, Sweden, Norway and Denmark liked hexagonal or octagonal types, while Germany itself favoured small corner paintings. The type made for England was usually circular and plain with a mahogany rim and a convex glass held by a polished brass bezel. No flower painting was included, but the movement did not differ from that of the usual type of Black Forest clock, which was also sold in England but on a much more limited scale than the type with circular dials. In England, Black Forest clocks were always called 'Dutch' clocks, a name possibly derived from the word *Deutsch* pronounced with a Black Forest accent by the travelling salesmen who disposed of them. Charles Dickens often mentions 'Dutch' clocks in his books.

A certain amount of elaboration went into Black Forest clocks from time to time. Dickens, in *The Cricket on the Hearth*, writes of a clock that possessed a figure of a haymaker who moved his scythe in time with the pendulum, and stood in front of a Moorish palace that contained the cuckoo. The museum at Fürtwangen has a clock with a haymaker but no Moorish palace or cuckoo, so we are left speculating whether Dickens based his description on a clock he had actually seen or whether he was combining several such clocks in his imagination. Moonphase is rare on Black Forest clocks, but day of the month is sometimes indicated, and less often, day of the week. Ting tang quarters are also sounded on some clocks, and sometimes the bells are struck by small jacks on a stage above the dial. Where much of the movement is made of wood, these features represent quite an achievement.

One of the greatest mysteries of the Black Forest is the so-called *Surrerwerk* (whizzing work) striking. Every time the clock strikes, the train runs long enough to sound twelve blows, even though only one blow may be sounded. The pin wheel that actuates the hammer has twelve pins on it which decrease in length from one to twelve. The hammer tail is moved automatically and is, in fact, controlled by the position of the hands, so that a desired number of pins shall actuate it when the clock strikes. Every time that striking takes place, the pin wheel makes one complete revolution, which absorbs a lot of the available weight fall and causes the striking work to need frequent winding. The only advantage of the system seems to be that the clock can be made to repeat, but as this can be accomplished by the ordinary rack and snail striking, one wonders why the Surrerwerk was ever made at all.

Some very small Black Forest clocks have been made which are known as *Schottenuhren*. The teeth of the wheels have to be more finely cut, but generally they are of the same design as the large ones. They are often provided with porcelain dials and sometimes have alarm work instead of striking.

The normal Black Forest movement with the striking train behind the going train, tends to sag after some years of use, and an improvement was effected in the general design by placing the trains side by side. This probably first occurred in the early nineteenth century. The earliest wooden clocks of the seventeenth century used stones for the weights, but when striking work was first introduced the weights were formed of glass tubes filled with sand, as two stones would have got in each other's way. The earliest striking clocks struck on glass bells, which was understandable in view of the already established glass industry, and even musical clocks with glass bells have been made. Later, when metal casting was done in the area, metal bells were used and later still the wire gong was used extensively for clocks to strike on, particularly cuckoo clocks. It seems traditional that the cuckoo is never considered capable of giving the number of the hour on his own, but it must always be sounded on a gong in addition. The only exception to the rule seems to be the small cuckoo clocks that have the bellows operated from the going train and sound every quarter of an hour.

Export trade from the area continued to flourish and expand during the early nineteenth century and everything seemed promising until in 1842 the American manufacturer Chauncey Jerome sent his first consignment of factory clocks to England. Jerome's clocks were of the O.G. pattern, rectangular cases about 26 in. by 15 in., a painted glass tablet in the door, weights enclosed entirely inside the case, and the clock capable of being stood on a shelf as well as being hung on the wall. The British Customs thought that Jerome's cargo had been deliberately undervalued, and exercised its right to take the consignment at invoice price plus ten per cent. Jerome was highly delighted and sent over another load which was dealt with in the same way. When the third load arrived, the Customs cleared it the usual way.

Jerome's clocks sold cheaply. They were often paid for on the instalment plan, and they were more sophisticated than the old Black Forest wall clock. The British public was won over and German sales in Britain began to decline. Jerome then began exporting to Europe and even to Germany itself, the previous home of the cheap clock.

The secret of Jerome's success was mass production. The idea had begun in America in the early years of the nineteenth century when Eli Terry began making hang up wall clocks with wooden wheels. Terry's first order was for four thousand, and people thought he was mad to make so many clocks as the market could not possibly absorb them. He was proved right, however, and others began to imitate him to the extent that there was no more demand for the product, and so he proceeded to design something fresh. This was a shelf clock, still with wooden wheels, but with the weights entirely enclosed in the case. He sold his old factory to some former employees and equipped a new one for producing the new clocks. Success followed his efforts and he began producing new and improved models of the clock. At this period Jerome was working for Terry, making cases and fitting movements into them.

A few years later, Jerome was in business for himself, manufacturing clocks with wooden wheels, but a financial panic in 1837 put an end to the wooden clock movement business and many manufacturers failed. Jerome then had the idea of a cheap brass movement to run one day. Previously, brass movement clocks had always been eight-day clocks, the one-day clocks having wooden movements, but Jerome developed his idea with the result that by 1842 he was in a position to begin exporting. The old wooden clocks had never been exported, as a journey over water would have affected the wooden mechanisms adversely. What Jerome had done others did also, and the export of clocks from America developed into a flood. Spring driven clocks were included in due course.

The American factory system had brought prices down and improved the product to a point where the old Black Forest system of hand labour and outworkers could no longer compete. A complete revolution was necessary if the Black Forest industry was to be preserved. Factory production of timepieces was begun in Schwenningen by Johannes Bürk in the eighteen-fifties, but these were mostly clocks for night watchmen and it was the Junghans family of Schramberg that took the first step towards the mass production of domestic clocks. Erhard Junghans was in business as a manufacturer of straw goods, particularly hats. His younger brother had been to America and worked in a clock factory there. He persuaded Erhard to abandon the straw business and the firm of Gebrüder Junghans was established to make clocks on the American system.

They began by making clock accessories and gradually swung into full production. Other firms followed suit and openly adopted titles in English and advertised the products as American clocks. There were for instance the Hamburg America Clock Company with the famous trade mark of crossed arrows, later amalgamated with Junghans, The Congress Clock Company, Concordia Clock Manufactory, Teutonia Clock Manufactory, Union Clock Company and so on. The movements were copied from the Americans and the paper pasted on the back of American clocks was also imitated. If an Eagle with Stars and Stripes could also be incorporated into the design, so much the better, and at least one maker adopted the American idea of illustrating his factory, with a horse drawing away a cartful of packing cases in front of it. After making the early American types, German factories branched out on their own lines, and by the end of the nineteenth century had re-established their export trade. It is a very interesting exercise to examine clocks of this period and try to decide whether they are German or American. The task is by no means easy.

By 1890 production in the Black Forest was reckoned in thousands of clocks daily. Fürtwangen contributed the greatest number, which included the products of the Union Clock Company, and the whole industry employed about fifteen to twenty thousand workers. The Technical School of Fürtwangen was established in 1877. The factory system in the Black Forest did away with a lot of sweated labour. Under the old outworker system, the men would work from 5 a.m. till 8 p.m. for a mere pittance while the middleman would live very comfortably by selling their products at the highest prices he could get.

The old type of Black Forest wall clock was not completely outmoded however. A new version was produced which abandoned striking work and used the space at the rear of the going train for alarm work. The arbors of the wheels were now of iron instead of wood, but basically the other features of the movement remained the same. The new type became known as the 'Postman's Alarm' and maintained steady sales in spite of the competition of American and German spring driven alarm clocks. In the eighteen-seventies the drum alarm with metal case had entered the field as a competitor, but the Postman's Alarm remained an article of commerce at least till 1914. Its success was probably due to the accuracy achieved by using a long pendulum and weight drive. The smaller wooden-cased spring alarms with short pendulums could gain during the early part of the day and compensate for it later on.

The cuckoo clocks produced after the change over to factory production were mostly weight driven, but a number of spring-driven examples were made, some of them even incorporating a quail to sound the quarters. While some had brass movements, others had the old type of wooden frame with brass bushes for the pivots, and a stopwork to prevent the springs being wound up too far which suggested the old stopwork used on table clocks and watches fitted with the 'Stackfreed'. Some of the spring-driven clocks even incorporated fusees. The 'Station House' type of case became popular about 1870 when railways began to be built in the district, and is still being made today although the present-day examples are brightly coloured, while the earlier clocks simply relied on the carving for the decoration. On a very large and expensive clock the carving could get completely out of hand, and feature nests with eggs, stags with spreading antlers and various types of birds and foliage. A smaller, plainer version of the cuckoo clock case was produced containing only a timepiece movement, and even as late as the nineteen-twenties the movements of this type were constructed of wood with brass bushes to take the pivots of the arbors. The weights of all types of 'Station House' clocks were invariably made to resemble fir cones.

The old Black Forest wall clock has now reached the period when it is beginning to take its place in the antique shops. It is reasonably robust, is capable of standing up to repairs, and it is to be hoped that it will soon be given the appreciation by collectors that it deserves.

PLATE 60   A Black Forest wall clock of the late 17th-century,
with a wooden movement and a 'Kuhschwanz' pendulum
before the dial. Quarter hour marks inside the hour figures are
indicated by a shorter hand, which virtually acts as a minute
hand.

PLATE 61   The movement of Plate 60 showing
wooden wheels and lantern pinions.

PLATE 62   A Black Forest clock of traditional type, less frequently found in Britain
than that shown in Plate 71.

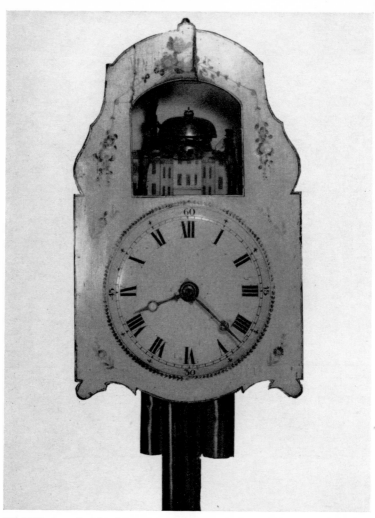

PLATE 63 A Black Forest wall clock, *c.* 1820, arranged to
strike ting tang quarters by means of little Jacks.

PLATE 64 A very early Black Forest movement in imitation of the American style *c*. 1870. The plates and wheels are cast and finished off roughly by a file. The back of the case bears a picture of the factory: 'J. Schultheiss & Sohn Black Forest Clock Manufactory.'

PLATE 65 Side view of the Cuckoo clock shown in Plate 66. Note the wooden arbors for the wheels and bellows above the movement.

PLATE 66 An early type Cuckoo clock, with the traditional Black Forest dial or 'shield'. 1800–20. Compare with the standard type of Black Forest wall clock (Plate 62).

PLATE 67 A Black Forest Cuckoo Clock, *c.* 1860, with pillars of white china. Compare the design of the case with Plate 76 and Plate 79.

PLATE 68   A Black Forest Cuckoo clock, *c.* 1860. The dog's
eyes move in time with the pendulum.

PLATE 69 A Trumpeter clock from the Black Forest (1870–80). This design of clock was a development of the Cuckoo clock.

PLATE 70 An early example of a Black Forest Cuckoo
clock in the form in which we generally know it today.
The extra door is for a quail which sounds the quarters.

PLATE 71 A Black Forest striking Wall clock, *c.* 1840. A typical example of the type made for the English market.

PLATE 72 A postman's alarm with a glass dial, *c*. 1880. The figures are painted on the rear of the glass and a wipe with a damp cloth keeps it clean.

PLATE 73   An eight-day Black Forest Wall clock movement
and the ultimate development of this type of clock. Date uncertain, may be 20th- century.

PLATE 74 A Black Forest Wall clock striking ting tang quarters, *c*. 1820. The arbors of the wheels are of wood, as are also those of the lifting pieces. Only two bells are fitted.

PLATE 76 A thirty-hour striking clock from Germany, *c.* 1880. Made by the Teutonia clock manufactory; one of the German firms with an American name.

PLATE 77 A Black Forest alarm, *c*. 1890. The movement has solid plates and is of good quality. Stopwork is provided to prevent overwinding. The case is veneered and is well polished.

PLATE 78 A German so-called 'Regulator'. Spring driven with imitation grid iron pendulum. This type of clock helped to bring about the end of the Dutch clock industry about 1880.

PLATE 79 A French Empire style clock of the pillar
and entablature design that became very popular at
this period. This may have been the inspiration of the
design of the clocks in Plates 67 and 76.

6

FRANCE AND FACTORY PRODUCTION

In the Renaissance period, France was one of the leading countries in Europe for clock production. Not only clocks, but also watches of fine quality were made, and many examples are now to be seen in museums all over the world. After the invention of the pendulum, France produced clocks of the Coster type which developed into the 'Religieuse', and by the early eighteenth century this had in its turn developed into a much larger clock decorated with ormolu or Buhl work and possessing glass sides and front so that the movement and the swinging pendulum could be seen. The pendulums often take the form of a sunburst, (cf 'Le Roi Soleil', Louis XIV). The Long Case clock, as we have seen, did not develop much except for the Comtoise type, and the finer examples by Paris makers are much rarer than their English counterparts.

The case of a clock in France was of far greater importance than in England, in fact French casemakers had to stamp their names in their products after 1750. The movement of a French clock always tended to be smaller than that of an English clock, and this was partly accomplished by the French makers usually omitting fusees which no English maker would have dreamed of doing. The early Coster clocks had going barrels, which showed great confidence in the new form of regulator, and the French makers in producing their 'Religieuses' followed suit. The earliest English clocks with pendulums that we can trace all have fusees, and it is doubtful if any were made in this country without them. France was much more reluctant than England to abandon the verge escapement in spite of the fact that the fusee was not popular with French makers. The very early English clocks have their winding holes very high on the dial, slightly above the centre line. The early French winding holes were usually symmetrically placed near VIII and IIII, but as the eighteenth century progressed, winding holes were placed in unsymmetrical positions.

This feature was in keeping with the changing design of the cases. The architectural features of the Coster type of clock gradually disappeared through the Louis XIV period and clocks acquired large amounts of ornamentation. Cases were decorated with Buhl or ormolu and the wood of the case was completely covered. The 'Religieuse' lost its straight lines and a waisted type of case became popular. Dials became smaller and were decorated with enamel plaques containing the figures, while English makers stuck tenaciously to the all-metal dial. Later, French makers acquired the art of making enamel dials on one piece, but all the time dials were getting smaller.

In 1730, Vernis Martin was introduced. This was an imitation of Oriental lacquer. An oak case would be veneered with pear-wood to give a better surface for the background colour under the lacquer. By the mid-eighteenth century, wood was being abandoned for clock cases except in the provinces and for Long Case clocks. New materials to take its place were white marble, bronze, and porcelain, and the latter would sometimes be made into vases which contained a movement, the clock indicating the time by means of rotating rings for hours and minutes which passed a fixed pointer, sometimes in the form of a snake. The Cartel clock became popular in France and also in Holland, Sweden and one or two other European countries. This consists of an unsymmetrically decorated case with decoration at the bottom which prevents its being stood on a flat surface, and it is therefore hung on the wall.

During the eighteenth century, France began to specialise in clocks which consisted of groups of statuary, and the movement was made smaller to interfere less with the composition of the group. It established the pattern for the French factory-made clock of the nineteenth century. The plates became circular and only slightly smaller than the dial, the springs were contained in going barrels and the teeth of the wheels were very finely cut. A comparatively heavy pendulum exercised control of the clock and made for good timekeeping. Although A. L. Breguet of Paris is usually thought of in connection with watches, his high standards of workmanship exercised a general influence on the clock production of the time. The French style of case also had its influence in England. One thinks of the products of Vulliamy at the end of the eighteenth and the beginning of the nineteenth centuries. The stage was being set for the extensive import of French clocks into England that took place during the latter part of the nineteenth century.

After the mid-eighteenth century, French styles began to be much simpler once more, and especially was this so after the Revolution. Straight lines and architectural features re-appeared, and some clocks during the Revolutionary period were made on the decimal system, although this never became really popular. When Napoleon was conducting his campaign in Egypt, Egyptian motifs found their way into French furniture and clocks.

During the First Empire, a style of case became popular in France consisting of four pillars on a base supporting an entablature with the circular movement suspended from it. The pendulum swung in the space between the pillars. This style was copied later in the century and also contained the germ of the idea for the thousands of marble and onyx cases that were made to house French movements as the century drew to its close. Whatever case was used, the circular movement with no projecting parts was adaptable and eminently suitable for factory production.

The French clock during the nineteenth century underwent a certain amount of simplification as far as the case was concerned.

Statutory designs

The idea of a group of statuary surrounding a rock in which the movement was placed gradually lost importance, and the central part containing the movement formed the foundation of the design, while the statuettes became appendages of it. This was probably due to the increase of factory production with an eye on ease of manufacture. A glass shade covering the clock was a necessity. Early illustrations of nineteenth-century French clocks are to be found in old editions of *David Copperfield* (1848) which also includes two rather vague drawings of the circular Black Forest wall clocks.

The glass shade was less popular in the later part of the century. By then the polished slate or marble cases had been produced and enjoyed great popularity. They fitted into the gloomy Victorian décor and enjoyed the advantage of being not easily movable, which preserved the clock from harm and made for more consistent timekeeping. So popular did they become that the American factories imitated them, fitting modified American type movements which were considerably larger than the dial and gave themselves away by having the winding holes in the chapter ring instead of considerably nearer the centre arbor. The Americans also imitated the visible escapement and provided the clocks with a pendulum that roughly resembled the French type. The marble case really reached the bottom of the social scale when it began to be fitted with balance-controlled timepiece movements of similar quality to the ordinary alarm clock.

A reversion to the idea of statuary was made in this period, for practically every marble clock was flanked by a pair of rearing horses. The French clock had come a long way from its eighteenth century delicacy, but in the centre of the group was still the precision-made circular movement that is probably one of the best high-quality productions on a commercial scale that has ever been made. These clocks are now being consigned to the scrap heap, as modern homes do not favour the strong mantelpieces necessary to support them, but it is a great pity that these movements have to perish as well. They tick quietly, keep good time, and go for years without attention. Lord Grimthorpe condemned French clocks, saying that the Black Forest types were better, but he was probably thinking of the silk suspensions which were still being produced in the nineteenth century nearly two hundred years after Huygens and Coster and the early Religieuses. He did however praise the type of French clock with a visible escapement and agate pallets, and admitted that the French were formerly better at turret clocks than the British.

Another French product of the nineteenth century is the carriage clock or *Pendule de Voyage*. The idea has long been held that a carriage clock was a clock fitted inside a carriage but, as the French name implies, it is simply a travelling clock. The movement, although rectangular, is constructed on the same lines as the pendulum-controlled circular movement just discussed, with the difference that it is controlled by a watch escapement and balance mounted on a separate platform. These movements bear a high finish as they are visible through the glass panels, and are attractive to modern eyes, in fact the carriage clock is one of the most popular clocks on the market today. They can be found in various grades. At the bottom of the scale we have the simple timepiece with a cylinder or lever escapement, then follows the same clock with alarm work, and the highest quality clocks are fitted with alarm work and striking, giving the opportunity for repetition at will. The Americans imitated the carriage clock also, using a very rough movement in place of the high quality French type.

Carriage clocks were mostly made in the rough in factories near Dieppe, and were sent to Paris to be finished. Some were eventually sold with the name of a vendor not resident in France shown on the dial, but the manufacture of the genuine carriage clock of quality is almost exclusively a French monopoly. The Comtoise clocks also nearly always bear the name of the vendor rather than the maker.

By the last quarter of the nineteenth century, France and Germany were both producing extensively on the factory system.

'English Dial' clocks

The French product was of higher quality than the German, but both enjoyed very wide sales and in particular exported large quantities to England. The great exhibition of 1851 drew the attention of the British public to what was being produced abroad and helped to encourage overseas trade, but the British horologists remained surprisingly blind to what was going on elsewhere. They continued to produce the type of clock that they were used to with solid plates and pillars, fusees, and generally rather large and heavy wheels, using the old methods which had sufficed for the previous two centuries. Tompion was one of the pioneers of quantity production, and kept a staff busy producing the types of clock with which his name will always be associated, but he was working for wealthy customers who could afford hand work and quality. A century and a half later the demand was much wider, and the public was consequently attracted to clocks produced in factories at a correspondingly lower cost. The inevitable result was that British clockmaking almost ceased, and the trade went to foreign factories that could produce much more cheaply.

One type of British clock did, however, continue to be produced. This was the 'English Dial', a timepiece with solid plates and pillars, a fusee and generally an anchor escapement with a comparatively heavy pendulum. The cases were generally of mahogany, and had a twelve- or fourteen-inch dial surrounded by a mahogany rim and brass bezel. The dials themselves were of painted sheet iron. These clocks are still familiar objects in shops and public halls, railway stations etc., although the tendency is now to replace them with clocks driven from the mains. Some of the earlier ones had very well-finished movements, and very early examples can be met with possessing verge escapements. Some are varied by having a longer pendulum with the case projected below the dial, the projection being either polished and left plain, or possessing a window through which the polished brass pendulum bob can be seen. Some early examples have the wood of the case decorated with brass inlay.

The same movement was also made with pierced and decorated

plates and known as a 'Skeleton Clock'. These are usually displayed under a glass shade, and a tradition has arisen that they were the work of apprentices who were making the clock to attain their mastership. While apprentices have made these clocks, there seems to be no justification for the assumption of a general rule. There is also a belief in some quarters that these clocks are French, but a French skeleton clock possesses a much less solid appearance and usually dispenses with the fusee.

A clock that enjoyed the popularity of the English Dial was bound to be imitated. German factories and also American manufacturers produced their own wall clocks with 12 in. dials which found a ready market in Britain. In particular, the Ansonia Company of U.S.A. produced a 12 in. striking dial with trunk and visible pendulum that enjoyed great popularity. Some 12 in. dials with French striking movements are met with but probably the movements only were imported and the cases made over here. The Black Forest version of the English Dial had a wooden frame with brass strips fixed in it to receive the pivot holes. In spite of this comparatively crude construction, the fusee was still included and the dial and bezel were an enlarged version of the older type of Black Forest wall clock which enjoyed such popularity on the British market before the American clocks arrived.

By about 1900 France and Germany were supplying the needs of most of Europe. A few English firms were active but only turned out a high-quality product. Some German factories produced a movement equal in quality to the French and found a market for it in England as well. They also produced their own versions of the Vienna Regulator, which perform quite satisfactorily, and evolved a spring-driven version somewhat shorter with a gridiron pendulum bearing the letters R/A on the enamelled bob. The letters stand for 'Retard' and 'Avance' and indicate the direction in which the rating nut below the bob has to be turned to produce the desired effect. This type of clock was usually known just as a 'Regulator' and was one of the main causes of the demise of the Dutch clock industry mentioned previously. 'Regulators' were made in two qualities, one with a movement approaching the finish of the usual French clock and another of a more American pattern with stamped-out plates and wire lifting pieces etc. Some factories even produced a very short version of the type. The 'Regulator' was produced in modified styles in the nineteen-twenties, usually having an oak case, bevelled edges to the glass panels, a pendulum with a wooden rod, and sometimes quarter chimes. Some even incorporated an idea that had been used in some of the clocks produced in Holland just after Coster's first pendulum clocks appeared, i.e. driving two trains from one mainspring.

Mention should be made of the four-hundred-day clock which was largely a product of the German factories and became popular in the late nineteenth century. The torsion pendulum was tried by Huygens after he had obtained his patent, in an effort to improve his invention, but it was not developed commercially until the mid-nineteenth century, firstly by Aaron D. Crane in America and later by various German factories. The type was popular up to 1914 then declined and had a revival after 1945, many new factories taking up production, but most of these have now disappeared and production is confined to one or two establishments. Features of the modern four-hundred-day clocks which the old ones did not always possess are levelling screws in the base, moonphase indicator, and a tube to carry the wire on which the pendulum bob is hung to protect it when being shifted.

As the twentieth century progressed, Germany came to the fore with mass produced quarter chime clocks, and plain hour and half hour striking clocks, mostly in cases the shape of Napoleon's hat. Long Case clocks were also produced, both spring and weight driven. A feature of clocks for the Continental market was the 'Bim Bam' strike where two different notes are sounded for each hour in place of the normal single blow. To have to listen to a clock striking twelve on this system can be irritating, especially as the striking is very slow. One or two British factories such as Enfield produced similar clocks to the Germans, but the total production was only a fraction of the import figure. After 1939 of course, clock production in Europe virtually ceased, and it was not until the post-war period that the industry reorganised itself and prepared to meet the demand caused by the general shortage of timepieces and changes in public taste.

PLATE 80   (*Left*) **A *Religieuse*, I. Thuret, Paris, *c*. 1690. Gilded chapter ring, black velvet ground to dial. Ebony case with Boulle decorations. (*Right*) A *Religieuse* by 'Marguerite à Paris,' *c*. 1675. The iron dial is covered with black silk. The case is ebony with tortoiseshell decoration.**

PLATE 81 *(Left)* A French *Religieuse*, Isaac Thuret, Paris 1670–80. Compare the general design with Plate 49. *Crown Copyright Science Museum, London.*

PLATE 82 *(Above)* A *Religieuse*, signed Gaudron, Paris, *c.* 1710. Every minute is numbered and separate enamel plaques are provided for the hour figures.

PLATE 83 This clock by Jacques Thuret (1694–1712) is veneered with Boulle marquetry. Jacques was the son of Isaac, who is represented in Plate 80 (*Left*) and Plate 81. Note the complete break-away from the Dutch model.

PLATE 84 (*Above*) **A Mantel clock by Ferdinand Berthoud,**
c. 1770.
(*Below*) **A Mantel clock with figures representing Night and**
Morning (after Michelangelo). The movement is by Jean
André Lepaute, 1709–87/9.

PLATE 85 *(Far Left)* **A pedestal clock. Boulle marquetry on oak possibly from Boulle's own workshop. The movement by Mynuel, Paris, working 1693–*c*. 1750. Sunburst pendulum bob. The hours are on separate plaques of enamel.**

(Right) **A Pedestal clock, 'Jacques Thuret Paris,' 1694–1712. It has a so-called thirteen piece dial and boulle marquetry on oak. The figures represent Love triumphing over Time. The clock and pedestal may not belong to each other, but a similar pair are found in the Louvre.**

PLATE 86 *(Right)* **This Pedestal clock is somewhat later than the other two (about 1750.) The figures are now being made more independently of the case. The movement is engraved 'Vidal à Paris.'**

PLATE 87 (*Left*) A musical clock in rococo form. The pendulum bob is in the form of Apollo on a sunburst. It contains 14 bells. The dial is inscribed 'Daillé Horloger de Madame la Dauphine,' the backplate is engraved 'Daillé à Paris' (working 1722–60). The musical portion is by Stollewerck, working 1746–75.

PLATE 88 (*Right*) A Cartel clock; case by C. Cressent, 1685–1768. The figures represent Love triumphing over Time. The dial is of one piece and may date from about 1758. The movement may date from before 1747 and is inscribed 'Guiot, Paris.'

PLATE 89 *(Left)* The Avignon clock. Bronze chased and gilt, the gilt being partly matt and partly burnished. Presented by the city of Avignon to the Marquis de Rochechouart, 1771. Movement by Nicolas Pierre Delunesy (1764–83), Paris.

PLATE 90 *(Right)* A Regulator clock. Case by B. Lieutaud, d. 1780, movement by Ferdinand Berthoud. The front of the case includes a Barometer. Compare the urn above with Plates 76 and 54 and note how small the dial is when compared with contemporary British examples.

158

PLATE 91 (*Left*) A French Mantel clock with moving hour and minute tracks. It has been suggested that the clock represents Louis XVI on his accession in 1774. The clock strikes hours and quarters.

PLATE 92 (*Below*) A French Mantel clock in white alabaster with gilt horse as decoration, Japy Frères, Paris, *c.* 1880. A typical French circular movement of a type exported in large quantities.

PLATE 93   A Mantel clock by Ferdinand Berthoud, late 18th-
century. Note the unsymmetrically placed winding holes.

PLATE 94 An English Dial clock, 'John Grant, Fleet Street', early 19th-century.

PLATE 95 An English Dial by John Benning, Windsor, 1780–90. Silvered and engraved dial, verge escapement.

PLATE 96   A German Mantel clock, *c*. 1900. This clock strikes ting tang quarters, and the full hour only at the hour. The movement is very solidly made. It represents a great advance on Plate 64.

7

CLOCKMAKING IN OTHER EUROPEAN COUNTRIES

Apart from France, Germany and Holland, we do not find other Continental European countries producing clocks in large quantities. Most of the other countries made a number of hand made clocks following first German or French models in pre-pendulum days, later copying English examples and turning towards French in the late eighteenth century. The cheaper markets in these countries were catered for by imports from the Black Forest.

The horizontal dial table clock lasted longer on the Continent than it did in England where it was seldom made, most of the examples being imported from France or Germany, but in other European countries we find it being made quite late in the seventeenth century, or even later still. For instance in Sweden it continued into the eighteenth century, even being fitted with two hands covered by a glass, and eventually developed into a large travelling watch with enamel dial.

The Long Case clock also appears in the latter country, the first examples being very English in appearance, but the hands, although pierced, did not possess the same finish that would be found on an English clock. After the mid-eighteenth century, French influence came to the fore with curves replacing straight lines, and the Swedish Long Case clock then possessed a case based on French models but generally less slender. The cases were usually light coloured and covered with painted designs relieved by a little carving. The enamel dials were reminiscent of France but generally larger, and the cases usually possessed a window to show the pendulum bob. The hands show French influence. By 1800 or so very plain cases had arrived which also correspond to French taste at that period.

The Bracket clock was also made in Sweden and followed the English model, but the finish was usually inferior to that of the prototype. As the eighteenth century progressed, the Swedish clocks differed from the traditional English shape and also bore more decoration. The carrying handles on the top were much more elaborate. Some Swedish clocks included the scalloped minute band usually associated with Dutch clocks. Decoration with lacquer is not unknown. By about 1800 French influence had really come to the fore, and we find small circular movements with cases of marble bearing metal decorations, and exposed sunburst pendulums. This type of case usually included a large number of pillars in the design.

The balloon shaped clocks popular in France under Louis XV were also made in Sweden, and the Cartel clock was copied in this country just as it was in Holland. The German Telleruhr type was also made in Sweden with very delicate hands, and a variation of this design was also mounted on a stand and provided with a revolving dial having the figures cut out of it so that it could be stood in front of a lamp and indicate the time at night.

About 1770, Arabic figures became popular in Sweden and were used on all types of clocks. They are often shown with 12 upright and 1 to 3 having an increasing amount of slope so that 3 becomes horizontal, and 4, 5 and 6 gradually become upright again with the performance repeated from 7 onwards. The names of the makers of Swedish clocks together with their places of business are usually boldly shown on the dial.

Denmark's clockmakers appear to have followed much the same path as those of Sweden, except that Long Case clocks tended to remain nearer English designs. Light coloured cases are known, but the painting found in Sweden is not so popular. The Long Cases of Danish clocks are usually decorated with more carving than those of their English prototypes. Danish makers have also produced various styles of table clock, and also the German type of Telleruhr.

The greatest name in Danish horological history is that of Urban Jurgensen who is mostly famous for watches and marine chronometers, and therefore scarcely comes within the scope of this book.

In Spain we find clocks of greatly contrasting types. There are Long Case clocks with chimes and musical movements of very high quality, and wall clocks approximating to the British Lantern type with movements almost as primitive as those of the fifteenth-century wall clocks. Sometimes the trains are arranged as on the French Comtoise clocks, and the bells usually stand very high above the movements.

Bracket clocks are also found, often with complications, but are generally late eighteenth-century types. It is possible that Spain may have imported the movements of these clocks, finished them and mounted them into locally made cases.

Italy, of course, was concerned with the beginnings of the mechanical clock. Milan claims the first public clock in the world and other Italian cities such as Padua were also early in the field. The astronomical clock by Dondi was a fantastic achievement for its time. Even the reproduction made in 1960 with modern tools is an object worthy of admiration, and how much more so is the original, being the work of only one man with very little previous knowledge of complicated clocks to draw upon.

The monastic alarm clock may have had its origin in Italy. An early one is preserved in a private collection in Bologna and another is illustrated in an Italian Tarsia panel in the Victoria and Albert Museum in London. Italy was well to the fore in the sixteenth century where Table clocks were concerned, and also watches of very high quality were produced by Italian makers. Wall clocks resembling the English Lantern type were made, and in the seventeenth century a dial indicating only six hours became fashionable. This had the advantage of retaining the simple mechanism for driving an hour hand only, but gave a closer approximation to the time than when twelve hours were included on the dial. A little thought was necessary to convert the reading of a six-hour dial to the twelve-hour system.

Some very fine cases have been made for Italian clocks. During the mid-seventeenth century it was fashionable to have clocks with lamps inside for night use. The figures would be cut out of the dial which rotated before the lamp, and the light shining through the figure would indicate the time. These were produced in London and South Germany as well as Italy, but did not have a very long life, as repeating work proved more popular from about 1680 onwards. The cases of these clocks tended to be large while the movements were small in order to accommodate the lamp. The Italian makers had graduated to small movements through making table clocks, while English makers still tended to make movements larger as their tradition had been more with the Lantern clock than the table clock. An English Long Case movement of the early period will have quite small wheels near the ends of the trains, but very large and heavy main wheels and barrels. On English spring-driven movements the fusees and barrels will be large and substantial, however fine the other wheels may be.

The style of wooden case used in Italy in the mid-seventeenth century is known as an Altar Clock. The term is a German one and is derived from the shape of altars at that time; it does not imply that the clock was intended to be placed on an altar. The type did not spread to England, to any great extent. An example by the famous London maker Edward East is known, and some English night clocks have features in common with Italian models of the time, but its influence on the English clock was small.

It did however, provide an interesting development on the island of Malta, and that about a century after it came into favour in southern Europe. These Maltese clocks date from the early nineteenth century, and were first introduced by a man called Kaladoniju Pisani in the village of Siggiewi; later other makers followed his example including those in the village of Zebbug. The earlier clocks possessed an hour hand only, but later ones were fitted with two hands. These clocks are extremely rare outside Malta as they are often heirlooms in Maltese families and very highly prized.

The Maltese clocks have a certain affinity with the American shelf clock in the design of the dial, and also in that the line for the weight is carried upwards over a pulley to give the maximum amount of fall. The movements are very small with the main wheel at the top and the escapement at the bottom with a tiny pendulum hanging straight down from the movement. This is typical of the Italian night clock. The cases are about 24 in. by 18 in. and decorated by paint and gilding suggesting Holland, Austria or any of the painted furniture found in various parts of Europe, while the cases of the Italian clocks are decorated by polished wood.

The dial and movement of a Maltese clock swing out on a hinge as on the Coster type of clock, and another feature in common with these clocks is the little aperture below the chapter ring to allow the pendulum bob to be started without swinging the movement out of the case. This also gives a visual indication that the clock is going.

The going period of the Maltese clock is usually only twelve hours and the earlier ones possessed a single hand only, which, however was artistically made, and provided with an extension that gave a balanced appearance. Little is generally known about Maltese clocks, but the type seems to have been created, had its day, and then passed out of fashion without having had a chance of developing. It is a pity that the Maltese clock never became an export commodity.

Before we leave the subject of Italian clockmaking, it should be noted that Italy very nearly pioneered the pendulum clock. The writings of Galileo and the clock by Treffler for the Medici Palace have already been mentioned, and the Science Museum also possesses a little table clock by the Italian maker Camerini, dated 1656, which has a short pendulum in front of the dial and does not appear to have been altered from balance control. No documentary evidence is available to authenticate the date of this clock, but if the date is correct, Italy had solved the problem of accurate time measurement before Holland did.

PLATE 97 (*Above*) A late 17th-century Table clock by Metzke of Sorrau. Note the curved minute track which was popular in Holland.

(*Below*) A horizontal Table clock with pierced case and single hand by Michael Brutscher, Stockholm, 1642–71.

PLATE 98 (*Left*) **Long Case clock by Johan Bastubak, black-smith in the parish of Kaarlela,** *c*. **1740. The outer edge of the chapter ring has four divisions between each hour, each of these divided into five parts.** (*Right*) **Long Case clock by Martin Widlund, master in Kokkola. The case is of pine, and the painted decoration is signed 1782. The general design of the case is reminiscent of English work about 1700.**

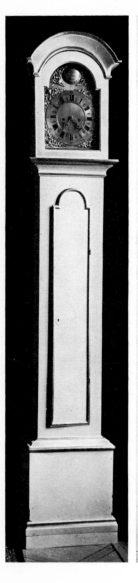

PLATE 99 (*Left*) A Long Case clock by Johan Forsell, master in Turku, 1746–66. The case is of pine. The white painting is not original. (*Right*) Long Case clock by an unknown Finnish maker. Dated 1784.

PLATE 100 (*Right*) A Swedish Long Case clock by George Haupt, Stockholm, late 18th-century. The case is more decorative than that of the typical English clock of the period.
(*Far Right*) 'A Farmhouse' clock as made in Sweden and Finland, late 18th- and early 19th-century.

PLATE 101 (*Right*) **A Finnish case for a 'Farm-house' clock.** (*Below*) **A Regulator by Otto Rautell, Helsingfors. Compare the heavy appearance of the pendulum with the lightness of the weight.**

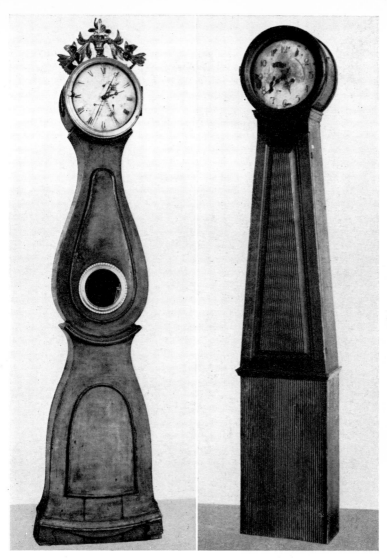

PLATE 102 *(Left)* A 'Farmhouse' type case with more complicated movement. *(Right)* An early 19th-century 'coffin' type case.

PLATE 103 (*Far Left*) **An early 19th-century Regulator clock, signed 'E. R. Wickman, Borgå (Porvoo), with gridiron pendulum.** (*Right*) **A Finnish regulator with indicators for the day of the week and the day of the month.**

PLATE 104 **A Long Case clock by Elias Ekblom, Turku (1830–55).**

PLATE 105   The dial of a 'Farmhouse' clock. This type
was made in Sweden and Finland during the late 18th-
and early 19th-centuries.

PLATE 106   The dial of a Regulator clock by Otto Rautell, Helsingfors. The hour indicator is arranged to rotate anti-clockwise.

PLATE 107   A late 18th-century Cartel clock from Finland,
also typical of contemporary Swedish work but derived from
a French model. Note the use of Arabic figures.

PLATE 108 (*Left*) **Clock on bracket by Petter Ernst, Stockholm, 1753–84. This clock is typical of French work.** (*Right*) **An Italian Altar clock of the mid-17th-century.**

PLATE 109 (*Far Left*) **A late 18th-century Norwegian Long Case clock.** It is based on an English design but the painting on the case is typically European. (*Centre*) **A country-made Norwegian Long Case clock.** (*Right*) **This Norwegian Long Case clock is inscribed 'Ole Larsen, Frohaug 1760.'** The general design of the case would suggest an English model somewhat later than this date and the painting suggests the lacquer decorations which were applied to many English Long Case clocks in the early 18th-century.

PLATE 110 (*Right*) **A Peasant clock from Upper Austria, dated 1563.** Note the iron movement, dial of carved wood, calendar and moonphase indicators.

183

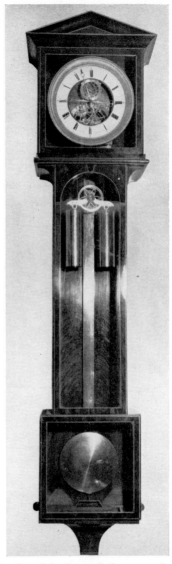

PLATE III *(Left)* **A Long Case clock with day of the month indicator, quarter and hour strike and carillon, signed 'Nicholas Lambert London'** (1750–70). *(Right)* **This Vienna Regulator goes for one year on winding. It is signed 'Franz Sterl, Mauer'** (1820–55).

PLATE 112   A Bracket clock signed 'Boschman London'. The numerous subsidiary dials are calibrated in German and the clock is probably of Continental make.

PLATE 113 (*Left*) **This Vienna
Regulator with Gothic style
case is only 85 cm high. The
going period is one month. It is
signed 'Josef Ratzerhof, Wien'**
(1838–58). (*Above*) **This little
clock is only one inch high,
and has a very rapidly swinging
pendulum in front of the dial.
It is signed 'Joh. Rettich,
Wien' (court clockmaker 1826–
71).**

Late the property of H.R.H. The Duke of Sussex K.G.

From a Model design'd
l. Prince Rupert.

PLATE 114   A Night Clock indicating the time by an illumi-
nated moving numeral showing the hour, which moves past
figures I, II, III to show the quarters. Mid-17th-century:
Joseph Knibb, London.

PLATE 115   A Maltese clock with single hand and quarter-hour marks inside the hour figures. Note the elaborately painted dial and slot to show the pendulum bob, also the decorated bracket and 'ear pieces'.

188

PLATE 116   A Maltese clock. Note the carving on the case
and the 'ear pieces' near the top. The slot shows if the
pendulum is swinging. Earlier examples had an hour hand
only.

8

CLOCKS AFTER 1945

1945 saw the French and German horological industries in a
state of chaos. In Germany, particularly, many factories had been
destroyed together with the necessary dies, jigs etc. needed to
produce the models previously in production. It was necessary
to rebuild the industries from scratch, and also get into production as quickly as possible, for a six-year delay had caused a very
serious shortage, and a dearth of repairers did not help the situa-
tion.

Public taste was changing and the stress was on smaller clocks
for the future. Mantelpieces were becoming smaller and were not
able to accommodate the larger clocks of the pre-war years. The
trend lately has been for watches to replace clocks, and many
clocks that are being made now more nearly approach watchwork
in their mechanism. Wooden cases are less popular, metal and
plastic taking their place, and even movements have undergone a
radical change in that the electric types, both battery and mains
operated are yearly becoming more popular.

The balance is more popular than the pendulum as the clock
can be made smaller, and does not have to be accurately levelled.
Many people object to the ticking of a clock, and a silent alarm
clock has been produced. This idea is not entirely new, because
clocks were produced in the eighteenth century with stretched
gut for pallets to achieve this end.

A bold innovation to overcome the noise of the tick and also
retain pendulum control with minimised wear in the train was
produced by the Horstmann Gear Company of Bath in the shape

of a clock with a magnetic escapement. Here the pendulum controls the rate at which the train runs purely by magnetic attraction, and there is no mechanical contact between the two. The power needed to drive the train is less than with a normal escapement.

The 'drum' alarm appeared in the eighteen-seventies and has remained popular ever since. A much smaller version with going train only appeared in the 'eighties and has also remained popular. About thirty years ago it was fashionable to make these with dials much larger than the movement, and to provide chromium-plated cases. Sometimes the dials would be square, semicircular or other fancy shapes, and these clocks often appeared as prizes at whist drives. The movements were balance controlled as were the drum alarms from which they were derived, and the escapements were of the type known as pin pallet lever. German manufacturers adapted this pin pallet movement to pendulum control, and produced little wall clocks in the shape of a mountain house with a tiny rapidly moving pendulum about 3 in. long. These clocks were first seen in Britain in the nineteen-thirties but are less popular here now, although they are still frequently met with on the continent. It is interesting to note that the pin pallet escapement had been developed for pendulum control by about 1900, but in a movement about the size of that of a normal drum alarm.

Alarms appeared originally in a plain metal case with the bell or bells on top, and through the years the case has appeared in different colours, shapes and materials, but some German factories are now reverting to the original shape with the bells visible above. Not only in alarm clocks but in other types of clock are older styles being repeated. The miniature Lantern clocks with balance-controlled movements have already been mentioned, and replicas of Bracket clocks are now being sold with similar movements. In Holland a style known as *Engelse Tafelklok* i.e. 'English Table Clock' or what we would call a Bracket clock is popular, but these clocks usually have pendulum movements (modern

German or French) and also striking work. They often incorporate moonphase and bear a fictitious maker's name on the dial such as 'John Smith, London'. Their style of finish is more typical of the Dutch version of the Bracket clock than the English.

In Germany too, antique styles are being reproduced for modern clocks. The Black Forest type of wall clock has its modern counterpart in eight-day weight-driven movements fitted with arch dials. Sometimes the dial is white with painted designs of flowers as on the old clocks, but sometimes it is of metal and suggests eighteenth-century clocks of higher quality. The pendulums are generally shorter than on the old Black Forest clocks and are more substantially made, while the weights are much larger and have brass cases.

After 1945, the British government decided to re-establish the horological industry, and factories have been opened in England, Scotland and Wales. The first priority was production of alarm clocks, but other styles came into production later. An English-made cuckoo clock came on to the market about 1950, and it possessed a movement of higher quality than that of the usual Black Forest cuckoo clock.

Contrary to the usual belief, Switzerland is not a great producer of clocks, preferring to specialise in watches, but Switzerland now produces small clocks with watch type movements either based on antique styles or in the modern idiom. Italy produces alarm clocks of its own design, and Junghans have opened a factory in that country for producing their own models.

Entirely new features found in clocks of the last twenty years are an adaption of the Chronometer escapement used in the 'Secticon' models, and the 'Floating Balance' which does away with pivot friction.

The revival of interest in antique styles for modern clocks is complemented by an ever-spreading interest in genuine antiques. During the last few years the prices of antique clocks have risen to fantastic heights, Tompion and the seventeenth-century London makers arousing particular interest in the sale rooms. While these

men and their productions have the main interest focused on
them, clocks by minor makers and even quite late clocks have
greatly increased in value in the last few years. There is also
growing up an educated public that can appreciate the finer points
of clock design and wishes its antique clocks to be restored in the
same style as the makers created them. Many Bracket clocks were
converted to anchor escapement in the nineteenth century, and
the present owners of these clocks are now having the verge
escapements reinstated. Lantern clocks which had their balances
removed in favour of pendulums perhaps two centuries ago are
now being converted to their original condition. The shortage
of skilled craftsmen to do this work means that the price of such a
conversion is high, and it may be anything up to a year before the
work can be carried out.

9

EUROPEAN INFLUENCE
OVERSEAS

Outside Britain and Europe, only two countries have created their own school of clockmaking, viz. Japan and U.S.A. Both countries began with European models, developed them according to their own ideas, and then settled down in the factory era to producing types that were also in production in European factories.

Japan had a very complicated system of time measurement prior to 1869 when western customs began to be adopted. The day and night were divided into a certain number of hours, which meant that the length of the hour during day and night varied throughout the year with long nights and short days in winter and vice versa.

The earliest Japanese clocks were based on the mediaeval weight-driven clocks or the English Lantern clock, and probably date from the seventeenth century. They possessed two foliots, one for day and one for night, and as day or night ended, the clock automatically switched to the other foliot, and therefore began to run more slowly or more rapidly according to the position of the weights on the new foliot. A visiting clockmaker would call periodically to adjust the weights and thereby give the clock a different rate as the seasons progressed. The Japanese house was usually too flimsy to support a clock from the wall, so these clocks were often kept on stands, which suggests some of the

Japanese and American clocks

Renaissance table clocks with dials on all sides that were placed on pedestals so that all dials could be easily inspected

Later Japanese clocks indicated the time by using the weight as a pointer to show the passing hours as it fell, the hours being mounted on separate plaques which could be moved nearer to or further from each other to adjust the length of the hour according to season. This simplified the movement, as the clock always ran at the same speed and therefore did not require its escapement or balances to be duplicated, and it could now be made smaller and more delicate and was light enough to hang on the central pillar of the house. After 1870, the European system of time measurement was introduced into Japan, and later clocks were produced in factories.

In the United States of America clocks can be divided roughly into three styles:

1 Colonial   2 Post Revolutionary   3 Factory

These types did not fall into three successive periods of history, for in the early nineteenth century all three types were being made at the same time. If we consider that the earliest type consisted of clocks of English design made by the earliest settlers, which developed along similar lines to English clocks, we find that the period of their making covers roughly the years between 1700 and 1840. The second period begins a few years after 1776 when the country was settling down after the War of Independence, and also lasts down to about 1840. The factory period begins about 1807 with Terry's earliest venture into mass production and has continued until the present day.

In the early eighteenth century most settlers in North America were British, and they continued to make the type of clock they had known in their home country, i.e. the Lantern clock and the eight-day Long Case clock. Sometimes the movements were very primitive as supplies of metal were difficult to come by, but the basic design was not very different. As the Birmingham factories began more and more to supply the British clockmakers with clocks which they sold as their own, an export trade to America was building up and clockmakers over there were supplied with movements in the same way. The cases in America were, of course, locally made, and often included different woods from those used by British casemakers. The style of case generally followed British designs, except that the trunk tended to be more slender and the hood more decorated.

So much for group 1. The real interest in the story begins after 1776 when American makers branched out on completely new designs, and at the same time incorporated ideas from other parts of Europe. This development was aided by European clockmakers who emigrated to the new country and took their traditional ideas of clockmaking with them. If one examines a list of American clockmakers, one cannot fail to notice the number of French and German sounding names about 1800, particularly the German. Another influence was the ordinary emigrant who took his household effects to America and included a European clock, which American clockmakers could see and use as a source of ideas for their later productions.

The earliest departure from the traditional British types was the 'Case on Case' clock. This appeared in the seventeen-eighties and was an attempt to produce a substitute for the spring driven clock of the day without the complication of springs and fusees. The appearance of the 'Case on Case' was that of a spring-driven clock standing on a small cabinet, but in reality both of these parts formed one unit. The origin of the type is not known, but can well be derived from the French fashion of mounting a clock on a pedestal so that the clock and pedestal appear to be independent units while in reality the pendulum is carried down into the pedestal. It should not be forgotten that France was one of the earliest countries to recognise the independence of the U.S.A. and even supplied military aid during the Revolutionary war, so it is not unreasonable to expect that French culture would make an impact on the U.S.A. at this period.

The Willard family working in the neighbourhood of Boston

were prominent clockmakers in the immediate post-revolutionary period, and in 1802 Simon Willard patented his improved time-piece, now known as a 'Banjo'. The design was the outcome of the desire to make a weight-driven clock even smaller than the 'Case on Case' type, and this was accomplished by having a heavy weight of limited fall acting on a barrel of small diameter driving a very large main wheel. In order to reduce the depth of the clock, the pendulum and crutch were brought to the front and worked in the space between movement and dial, the pendulum having a stirrup to embrace the motion work. This position for the pendulum was not unknown at the time as it was employed on the French 'Comtoise' long-pendulum clocks, some of which had no doubt been imported into America by 1800.

The later developments of the 'Banjo' viz. the 'Lyre' and the 'Girandole' seem to have an affinity with the Swedish 'Farm House' clocks which were made about 1800, although the latter were meant to stand on the floor and were much larger. Another feature common to both styles is the use of Arabic figures which also became popular on the early Terry clocks. In view of the popularity of Arabic figures on late eighteenth-century Swedish clocks, it is possible that the American fashion originated in Sweden.

Early in the nineteenth century some very plain Long Case clocks were made in Sweden, and the type also appeared in America where it was nicknamed a 'Coffin' clock. Possibly both types were traceable back to the revival of architectural styles in France in the late eighteenth century, especially after the Revolution.

The dial of the 'Banjo' is somewhat marred by having the winding hole opposite figure 2, so that the weight has a maximum amount of fall. Unsymmetrically placed winding holes have their precedent, however, in some of the clocks produced by seventeenth-century London makers and especially in mid-eighteenth century French clocks. The unsymmetrical placing of the winding holes became very common in American spring clocks of the mid-nineteenth century, but is no worse than on many French clocks made a hundred years previously.

The next important event in American clockmaking was the mass production of wooden clock movements by Eli Terry. These clocks were of the 'Hang Up' type and combined the wooden wheels of the Black Forest clock with the general arrangement of the English Long Case clock movement. The importance of Terry's venture into mass production was that its success encouraged him to produce his various shelf clock designs, which in its turn encouraged others to do likewise and so led to the establishment of clockmaking as an industry in Connecticut. It should not be forgotten that while new types of clock were being introduced into America in the early nineteenth century, many makers were still producing the conventional type of Long Case clock for which they relied on material supplied by English factories. The Jefferson Embargo of 1807–9 and the war of 1812–14 by stopping the import of this material helped to encourage the demand for the new types of clock.

Terry's shelf clock was made in several models, and in all of them the weight lines were carried over pulleys in the top of the case to give maximum fall, and this feature became well known on American clocks in later years. There is in the museum in Vienna a clock which possesses it, and is also housed in a style of case which became popular in America in 1820–30. The Vienna clock was made in the early seventeenth century.

Terry placed his movement well back in the case, and not only the pendulum was brought behind the dial but the escapement and locking plate were put there as well, while the motion work was between the plates. As far as can be ascertained this was an entirely original idea, but on one of the early Terry shelf clock models the pendulum was hung in front of the dial as on Black Forest clocks in the eighteenth century.

An important feature of the Terry clock has its origin back in the seventeenth century, viz. that the minute marks are inside the hour figures. The chiming clock by Nicholas Vallin in the Ilbert

Collection has a quarter hour dial inside the hour dial served by a hand shorter than the hour hand. This feature can also be found on eighteenth-century Black Forest clocks, and on many public clocks in Austria. The Terry hands were almost the same length, but were distinguished from each other by the hour hand having a larger design than the minute hand.

The lower part of the case of the Terry clock was filled by a glass tablet bearing a painted design, but a small part of the glass was left plain to allow the swing of the pendulum bob to be seen. This idea dates back to the 'Zaanse' clocks, for when the pendulum was in its infancy the owner of a clock would need reassuring that the clock was still going. In the late seventeenth century also the English Long Case clocks had a bullseye glass let into the door for the same reason, and the early English spring clocks with verge escapements had a dummy pendulum moving behind a slot in the dial. While the bullseye died out in the Long Case clocks, the dummy pendulum of the spring clocks remained in fashion as long as the verge escapement was used, for this type of clock was intended to be moved about, and the owner would require satisfying that the clock was going in each new position it occupied. The Terry hang-up clocks had, of course, exposed pendulums, but we find him making provision for allowing the pendulum bob to be seen in all his shelf clock models. As the wooden movement shelf clock became generally accepted, the fashion of leaving the plain piece of glass died out, to come in again when the one-day brass movement O.G. clocks were introduced. Once this type had established itself, the practice of painting the glass all over was resumed.

In the eighteen-twenties it became fashionable to make shelf clocks much taller than the Terry type and sometimes the dial would be banished right to the top of the case and the front of the clock formed of a looking glass. This type is particularly associated with New Hampshire. These clocks usually had brass movements and sometimes included the so-called 'Rat Trap' striking mechanism. This does away with the normal Locking Plate and instead has a series of holes drilled at increasing intervals in the rim of the great wheel. A wire taps on the face of the wheel for each blow sounded, and when it meets a hole the train is locked. There is a clock in the castle of Wurtzburg, Germany, which has its striking controlled by a lever tapping on the face of the locking plate instead of riding on its edge, and this clock has a feature which became better known in the later American factory-made clocks. Instead of the locking plate having a continuous motion, it is moved a short distance after each blow is struck, and remains stationary until the next blow. This feature is also found on the turret clock from the Hague in the Utrecht Museum, and on an old turret clock formerly above one of the city gates at Rothenburg, Germany.

The O.G. Case used by Chauncey Jerome for his one-day brass movement clock was probably derived from a German type of clock that had a case shaped like a heavy picture frame. The brass movement clock was eventually produced in eight-day form, and while some of them retained the O.G. type of case, others had a case with an entablature and two columns as in the French Empire style of mantel clock.

When American factories started producing spring-driven clocks in the eighteen-forties, many different styles of case were produced, among them the so-called 'Round Gothic' which is now known to collectors as the 'Beehive'. This bears a great resemblance to the 'Lancet' case of English spring clocks made in the Regency period. Another 'Round Gothic' style was produced by the firm of Ingraham in which the top of the lancet was brought to a sharper point supported by curves of small radius. This outline has been used for dials of wall clocks of the sixteenth-century.

Many American spring clocks are fitted with alarm work set by means of a disc mounted friction tight on the hour hand pipe. The dial is completely cut away behind this disc, a feature which also dates back to the early iron wall clocks. Some American alarm clocks have the alarm train between plates of its own forming a small unit quite independent from the going train. This feature can be found in some clocks of the Coster type.

English clocks were generally made to strike hours only, while Continental clocks usually struck half hours as well. Up to the middle of the nineteenth century, American clocks generally followed the English practice of striking hours only, but after this, some makers adopted the plan of fitting an extra tail to the hammer which was raised by the going train and allowed to fall at the half hour, thereby giving an aural indication of the time without the striking train being released. There is a sixteenth-century Gothic clock in the Science Museum, formerly in the Webster Collection, which also embodies this feature.

Late in the nineteenth century, American clocks were made without the painted tablets which had been so common in the earlier years and had plain glass fitted in their doors. The pendulum, now being visible, was often decorated with a large piece of thin brass bearing an embossed design which was similar on a smaller scale to the elaborate embossed pendulums fitted to the Comtoise clocks.

PLATE 117 *(Left)* A musical clock by Nicholas Vallin, 1598, a Fleming working in London. The quarter hour marks are inside the hour marks and are indicated by the shorter hand.

PLATE 118 An American O. G. weight-driven clock. E. N. Welch, Forestville, Conn., *c.* 1870. This type began the export to Europe and eventually drove the Black Forest Wall clocks off the market.

PLATE 119 A very small Rosewood clock; the Gothic style by Daniel Desbois, London, *c*. 1845. A similar style appeared in America at about this time.

PLATE 120 Long Case clock by Otto Edvard Fock, Turku (1823–68). The case is of birch, stained to imitate mahogany. Signed 'O. E. Fock, Åbo'.

PLATE 121 A small Black Forest alarm, in imitation of American models, *c.* 1875. The American type of open plate is used, and the escapement is between the movement and the dial. The pendulum's swing is visible.

PLATE 122   A Gothic style clock by James Gowland, London,
? 1820. Rosewood inlaid with brass, it is a possible anticipation
of the American Sharp Gothic.

PLATE 123   An English Long Case movement under construction. Right: chapter ring, movement plate and dial plate;
centre: various parts in semi-finished state; left: hands,
spandrels, winding key and weight pulleys.

IO

THE CRAFT OF THE CLOCKMAKER

It is often remarked in connection with old clocks how wonderful it is that they could be made without proper tools. It is of course impossible to make a clock without proper tools, and even the very earliest clockmakers must have possessed a fair collection, primitive though they may have been. Whatever period of history one studies it is generally found that technology at that time had reached a higher level than might have been expected.

The tools of the clockmaker's craft are almost as fascinating as the clocks they produced. One is constantly filled with admiration for the men who devised appliances for doing apparently impossible tasks, and it should not be forgotten that many of the old clockmakers made their own tools. It was, however, possible to obtain tools ready made, for instance John Wyke of Liverpool issued a catalogue about 1770 containing a far greater range of equipment than could be purchased today. In those days there was no mail order service for spare parts and no standardisation. If a part for a watch or clock was needed it had to be made on the spot.

The clocks of the mediaeval period would have been made with the tools used by the average blacksmith. Wrought-iron work had reached a high standard by the fourteenth century as can be seen from the railings round mediaeval tombs in various cathedrals, but while the normal blacksmith's equipment would provide most of the clockmaker's needs there were one or two special items that would have been necessary in addition, particularly in the making of wheels and mounting them on their arbors. To make a wheel, the mediaeval clockmaker would beat out a circle of iron, fasten in the spokes or 'crossings', and then mark out the teeth. The latter would have been laboriously cut out with a file and then finished with a smoother file, but after that came the problem of making a hole dead centre so that the wheel ran true. Today this would not be a great problem, but then it would probably have been done by trial and error, making a very small hole at first and enlarging it by filing until the wheel was near true as possible. Then would come the problem of mounting it on its arbor. The latter would have been produced from a bar of iron by hammering, and would require the pivots and wheel seating to be turned. Some primitive form of lathe would be necessary for this, probably at this period worked with intermittent rather than continuous motion. The groups of itinerant clockmakers would probably have had a boy for this purpose, whose other duties would have been to fetch and carry and work the bellows for the fire.

Pinions in the early days were of the lantern type. These were easier to make than the solid type and are also more tolerant of inaccurate depths. This would involve drilling holes in iron plates, turning shoulders on the leaves and then assembling the pinion by forging. Turning and drilling would require sharp tools of good steel. Steel making was a very uncertain process in these days, and only small quantities of variable quality could be produced. Frequent sharpening of the tool would have been necessary and as the tool wore down, different parts of it would have kept their edge better than others.

Drilling would involve the use of sharp punches to mark the position of the holes, and diamond-shaped drills to do the actual boring. A to and fro motion was produced by a bow, the string of which was coiled round a pulley mounted on the drill. The top of the drill would be supported by running it in a hole in a piece of iron held by the operator. Smaller holes could be drilled by one operator but the larger holes would require two, one to support the drill and the other to supply the motion. Another form of drillstock consisted of a rod with a means of fastening the drill in its end and a small flywheel just above this point. A horizontal piece of wood with a hole in its centre was then placed over the rod and each end of the wood was connected to the top of the rod by a cord. When the cords had been wound round the rod a few times, an up and down motion of the wood would cause the drill to rotate to and fro with the assistance of the flywheel.

After the wheels of the clock had all been cut and mounted on their arbors, it would be necessary to drill the holes to receive the pivots in the upright bars of the frame. It would be necessary to bore holes exactly opposite each other so that each arbor was dead square to the frame.

As clocks became smaller, the file became the all-important tool. Smaller parts required making with greater accuracy, and very fine files would be needed for tooth cutting and shaping the other delicate parts of the clock. The craft of the clockmaker began to have more affinity with that of the locksmith and less with that of the blacksmith.

The advent of brass into clockwork in the sixteenth century not only made for easier working, but also minimised friction in the clock itself, for there is less friction between parts made of different metals than between parts made of the same metal. Brass wheels working against steel pinions cause the steel pinions to wear first, as tiny fragments become embedded in the softer brass which then acts as an abrasive. The only known way of hardening brass is by pressure, usually by hammering. All brass used in clocks should be hardened, and in early workshops this was usually done, but in Holland and the Black Forest, wheels were cast and then had the teeth cut without any hardening taking place. A blow hole in a casting could ruin a wheel, and a finished wheel cut directly from a casting would be softer and less durable, but assuming that the process of casting was well carried out the number of failures would not be large, and the process would be generally cheaper. The better type of clock would have had the metal for the plates and wheels hand hammered before it was worked on. Special smooth-faced hammers were used for this purpose and the process is known as planishing.

The most fascinating and the most difficult operation in connection with making a clock is the cutting of the teeth in the wheels. The early clocks must have had their teeth marked out and laboriously cut by hand. One often finds punch marks on the wheels of old clocks indicating where the cuts between the teeth should come, and often there are signs that the work was not performed as accurately as it could have been. When one considers the table clocks and watches of the sixteenth century, the cutting of the teeth is a masterpiece of patience and skill. A record of a device for marking wheels has come down to us in the inventory of the tools of a watchmaker in Leeuwaarden made in 1600. This is a brass dividing plate with a ruler belonging to it that may or may not have had a guide for a slitting file. Unfortunately this apparatus is only mentioned in the document and we do not possess a drawing of it. There are two observatory clocks in the Hessisches Landesmuseum at Kassel in which the teeth are so finely cut that one cannot help believing that some mechanical means was used to perform the task. These clocks were made about 1600 by Jost Burgi, and are outstanding examples of delicate workmanship far ahead of their time. It is possible that in an ordinary workshop some form of jig was used for cutting teeth, but very probably the workman would mark out his wheel and patiently file away until it was finished. After the teeth have been cut, they are square in shape and have to be rounded off with a special file having teeth on one face only and with the back curved away from the edges and made smooth to avoid damaging adjacent teeth. This operation when performed by hand takes as much patience as the actual cutting.

The invention of the wheel-cutting engine is credited to Robert Hooke, and is believed to date from about 1670. It is probable that a more primitive form of this apparatus existed before this time, at least on the Continent. The blank to be cut is very firmly fixed to a vertical arbor which passes through the centre of a brass disc about 18 in. in diameter. This disc bears a number of concentric circles, and each circle has a number of equally spaced depressions in it, the number of depressions in the circles being numbers of teeth often required by clockmakers. If, for instance, it is required to cut a wheel of sixty teeth, the blank is mounted and the circle of sixty depressions chosen. An adjustable locking piece attached to the frame is secured in one of the depressions, and after the first cut is made the disc is rotated one division and the locking piece allowed to fall into the next depression in the circle. In this way, the blank is moved through one rotation in sixty equal steps. The cutter is in the form of a disc about the size of a penny with file-like teeth cut on its edge, and it is mounted in a small frame whose position can be adjusted so that the cutter passes through the blank giving a convenient depth of tooth. The cutter is rotated by means of a crank through gears or a belt drive, and as the blank is brought into each new position the cutter frame is depressed while the cutter is rotating rapidly and cuts one space on the blank to the accompaniment of a great deal of noise. An experienced workman can perform the task quite speedily. Some cutters are made so that they round up the teeth as they cut them, but where an ordinary cutter is used the teeth must be rounded up by hand as previously described. After the teeth have been rounded up, the wheel is 'crossed out', i.e. spaces are cut out of the centre for lightness, leaving spokes or crossings.

The Frisian clockmaker cut his wheels by means of a foot lathe. The cutters were made about 2 in. to 3 in. in diameter and mounted on wooden arbors, in much the same way as the wheels of the Black Forest clocks, with a metal bearing at each end. The cutters were mounted between the centres of the lathe and the blank to be cut was mounted on a small platform which allowed the blank to be advanced towards the cutter. The wheel had to be marked first, and the system was a compromise between the hand method and the cutting engine.

The Black Forest workman also used his lathe as a means of rotating the cutter, but he employed the division plate mounted on a hinged frame in front of it carrying the blank in a similar fashion to the wheel-cutting engine previously described. As each tooth was cut, the wheel blank was moved away from the cutter by rocking the whole arrangement on its hinge and then advanced the space of one tooth and moved towards the cutter again. This is virtually the same process as employed in the use of the wheel-cutting engine, except that the extra mechanism for rotating the cutters is dispensed with. The Germans call the apparatus 'Der Zahnstuhl'.

The making of the division plates was a highly skilled task. Most of the numbers required would be divisible by several factors such as 48, 60, 72 etc., but in the days of verge escapements a series of odd numbered circles would be required as well. Some division plates include the special spacing required for making locking plates. Division plates have been made of very large diameters to get greater accuracy, but the convenient size for the average clockmaker was about 18 in. to 24 in.

Turning is an important process in a clockmaker's workshop but the clockmaker of the past did not have elaborate lathes at his disposal such as are in use today. Most of the clockmaker's turning used to be done between dead centres (a 'Turns'), and the work was given a to and fro motion with a bow as mentioned earlier in connection with drilling. Sometimes a continuous rotary motion could be supplied by a hand or a foot wheel, and the apparatus is then known as a 'Throw'. The term 'Lathe' is only strictly correct where a continuously rotating spindle is provided for holding the work. The lathe came late into British clockmaking but was in use earlier on the Continent. The Frisian clockmakers drove their lathes by a foot pedal operating a flywheel supported from the ceiling which was connected to the lathe by a belt, while the Black Forest clockmakers used a flywheel driven by foot power placed under the bench. This flywheel often consisted of a grind-stone, so when it was desired to sharpen a tool, the clockmaker merely removed the driving belt while he performed the operation of sharpening and replaced it again afterwards. The Frisian lathe was also used for wheel cutting as previously described.

The tools used for turning were called gravers and consisted of bars of steel of square section ground off at an angle across one corner so that a lozenge shaped face was left. The size of the steel would be anything from $\frac{1}{16}$ in. to $\frac{1}{2}$ in. square. Sometimes a piece of steel of lozenge shape section would be used to begin with if it were desired to make a graver for cutting very deeply. Provided the gravers were kept sharp, practically any turning operation could be performed with them. The secret of good work was to use the part of the edge just behind the point. When the tool was held at the correct angle, the metal would come away in shavings rather than in dust. An adjustable tool rest supported the graver as near to the work as possible, and when intermittent motion was applied to the work, the graver was tilted away from the work slightly on the back stroke so that it did not rub the work when it was not being used for cutting.

Pinions were formerly made from pinion wire which came in lengths of about one foot. A rod of steel was forced through a series of dies, each bearing a more deeply indented pattern of a pinion, with the final one giving the full depth of the leaves. When it was desired to make a pinion, a clockmaker would cut off a suitable length of pinion wire with the correct number and size of leaves just slightly longer than the overall distance between the outside edges of the plates. He would then mark the positions for the operative part of the pinion and the wheel seating and then file each end of the piece of wire to a cone. The piece would then be mounted between female centres in the turns or throw and a groove cut across the leaves down to the solid part of the arbor to divide the pinion proper and the wheel seat from the rest. The piece would then be removed from the throw or turns and the unwanted leaves cut off. It would then be remounted, turned smooth, have the pivots turned on it and after all other work was done the ends of the pivots would be rounded by mounting the pivot in a special holder that allowed the end to protrude. The turns or throw always needed the work to be supported at both ends, unlike a lathe which can support one end only in a chuck. The leaves were finished by a file, hardened and tempered, polished and then the wheel was fastened to its seating. A modern pinion would be made by turning from the solid and cutting the leaves on the relevant part only. A similar method would have been used in the days before pinion wire was made, but would have been a very laborious process without the use of machinery.

The earliest clocks used lantern pinions made up of rods between two plates forming a kind of cage. The earliest Black Forest foliot clocks also used them, made up of wires between wooden discs, but the making of small lantern pinions by hand is a difficult job. The Black Forest continued to use them throughout the period when wooden-framed wall clocks were being produced there, for the wooden arbors used in these clocks lent themselves to receiving the ends of the wires. After factory production of clocks had been established in Germany, the German factories employed children to assemble them with an adverse effect on their eyesight. At this period, of course, the ends of the pinions were being made of metal. Arthur Junghans, of the firm of the same name devised a method of producing them mechanically. The arbors were fitted with the pinion ends, having the holes for the wires opposite each other, and were mounted in a metal box containing a large number of the short pieces of the wire. The lid was then closed and the box violently agitated causing the wires to fly in all directions, and at the same time the air was exhausted. After a short time each pinion would have the correct number of wires in place. Twelve of these machines could do the work of five hundred children and only needed loading by unskilled labour.

Wheel cutting under factory conditions is always done en bloc. A pile of blanks are clamped together, and a cutter moving along the pile cuts one slot and rounds the ends of the adjacent teeth in one operation. Wheels for motion work are sometimes stamped out complete in one operation, but wheels which have to transmit power are always cut. The American factories not only cut their wheels in batches but also used the same size wheels for the going and striking trains. This meant that far fewer machines had to be set up.

After pivots had been finished and smoothed they were burnished. A burnisher is a piece of steel like a file made without teeth and polished. When a pivot is rotating, the burnisher is moistened with a suitable lubricant and pressed on the rotating pivot with a to and fro motion, leaving a very smooth surface which is also slightly hardened. This makes for less friction and more durability when the clock is running. As far as the lubricant was concerned some workmen used saliva and other workshops would save the dregs of the beer cans for this purpose.

The simplest way to make a verge is to cut the staff with the two pallets out of a flat sheet of metal and afterwards warm the portion between the pallets and twist it until the pallets take up a position at 90° to each other. The verge is then finished in the throw or turns with a steady to hold it in the middle. This is a piece of steel fastened to the frame with a V-shaped slot in which the work fits. When carefully made, all signs of the twisting have disappeared by the time the verge is finished. The anchor for the escapement of the same name is produced from a small forging, and after the hole for the arbor has been drilled the rough anchor is set on a piece of paper with the scapewheel and carefully marked, after which the surfaces of the pallets are filed to shape and then smoothed. The anchor itself is then smoothed all over and hardened by being made red hot and quenched quickly in water.

When the wheels of a hand-made clock had been cut and mounted on their arbors, they were ready for mounting in the frame. The clockmaker would have marked the positions for the holes by careful measurement, in fact one often finds movements where the wheels and their centres have been set out on the front plate. In later days the depthing tool was available. This was like two sets of turns fastened together by a hinge at the bottom. Their distance apart was regulated by a screw, and the wheel and the pinion that it was required to drive were each mounted between a pair of centres, one on each half. They were rotated by the fingers while meshing together, and the position in which they ran most smoothly was the correct depth. The centres holding the arbors were produced at the ends of the tool and terminated in sharp points, which were used as divider points for marking the plates. When the plates had been marked, the pivot holes were started by a sharp punch and the holes drilled with a drill that was smaller than the pivot. The holes were then opened out to correct size by means of a broach, a tapered steel rod of hexagonal section which not only cut the metal away but hardened the inside of the hole. When the hole was almost its correct size, the cutting broach would be replaced by a smooth broach which would enlarge the hole only a very tiny amount but act in the same way as a burnisher and leave the inside of the hole very smooth and hard.

Such sophisticated methods did not, of course, commend themselves to the Black Forest makers who inserted brass bushes into the wooden frames to take the pivots. The bushes themselves were usually made of brass strip rolled into a circle. The depthing in a Black Forest clock is not, however, of great importance as a lot of latitude is necessary owing to the wood shrinking by age and the effect of moisture etc.

The clockmakers in this area adopted a tidy method of keeping their tools in circular racks hung above the bench something like a chandelier. One sometimes finds similar racks in offices for holding rubber stamps, but the office type are arranged to stand on a desk. The making of fusees was undertaken by means of tools based on the pantograph. As the rough fusee was rotated, a cutter was moved by rods which worked over a pattern and caused the cutter to trace out the correct curve. A similar tool could be used for screwcutting, but most makers formed their threads by a screwplate.

The making of chains is a tedious process. Pre-cut lengths of wire are laboriously bent into the form of the links by pliers and are strung together. It is possible to speed the process a little by winding wire round a piece of wood of rectangular section that fits inside a link, the wood having a longitudinal groove cut on one of its flat sides. The wire is then sawn where it crosses the groove and the links are fastened together by pliers. The Black Forest produced a chain-making machine which was worked by a hand crank and produced one link for each revolution of the crank, but it came rather too late in history to help the Black Forest industry very much. Dutch chains usually have the links in figure 8 form with the loops at right angles. They are made by means of two pairs of pliers, one with flat noses with a longitudinal groove down the centre, and the other with round noses. A pre-cut piece of wire is held by the grooved pliers with the wire in the groove and half its length protruding. A loop is formed by means of the round-nosed pliers and the loop is then held by the flat of the holding pliers and the other loop formed at right angles to the first. When a few links have been made they are strung together and the loops finally closed. Dutch chain is still made by hand today and is still sold by the 'El' (69 cm).

The making of dials is a separate trade from that of the clock-maker. The earliest turret clocks would have had their dials produced by painters who either painted directly on to the stone-work of the tower or on to a prepared surface made of wood or metal. Many of the oldest dials had a small roof over them for protection. With the coming of the domestic clock, the painter would again be responsible for the dials of the early weight-driven wall clocks, while the engraver would be responsible for the dials of the table clocks, and later for the Long Case and Bracket varieties. In a large city like London, the master clock-makers could call on the services of one or more engravers, but in the smaller towns it would often be necessary for the clockmaker to do his own engraving. By the time of the factory production of the late eighteenth and early nineteenth centuries the painter had come into his own again, decorating the sheet-iron dials with landscape or other designs and producing the figures by means of stencils. The iron dials of Dutch clocks were always hand painted, and one or two dial painters are still in business in Holland. The production of enamel dials and other enamel wares was always a specialised process divorced from clockmaking.

Wooden dials were always popular in the Black Forest and the early American factory clocks had them also, although a change over was made later to sheet zinc which allowed the dials to be printed. The Black Forest factories then took up the sheet zinc dial as well, and eventually paper and cardboard dials appeared on the cheapest clocks such as the well known drum alarm. Junghans were early in the field with luminous dials after the discovery of Radium in 1898.

Such then is the broad outline of the processes involved in making a clock. It is not possible to go into every detail of the construction, for this is not a technical book and the technically minded reader can follow the subject in more detail from the other works listed in the appendix. Sufficient has been said, however, to give the reader an insight into the hours of hammering, filing, turning, smoothing and polishing that went into a hand-made clock, and when we consider that the clockmakers of the past worked from dawn till late evening six days a week, and that it took a fortnight to produce an eight-day Long Case movement, our respect for such a machine should be very much greater. The factory-made clocks of the previous century are not without interest, for in the early days the proprietor was far more closely connected with the day to day production. Even a modern factory clock can have interesting features. The Ilbert collection included a silent ticking alarm clock made about 1938.

PLATE 124 Drilling by means of a bow. The string rotates the pulley to and fro, and the pulley is supported by a steel point in its centre which runs in a small hole in the jaw of the vice.

PLATE 125 An 18th-century wheel-cutting engine still in daily use. The dividing plate with the series of holes can be clearly seen.

PLATE 126 A 17th-century wheel cutting engine with crank
drive. *Crown Copyright. Science Museum, London.*

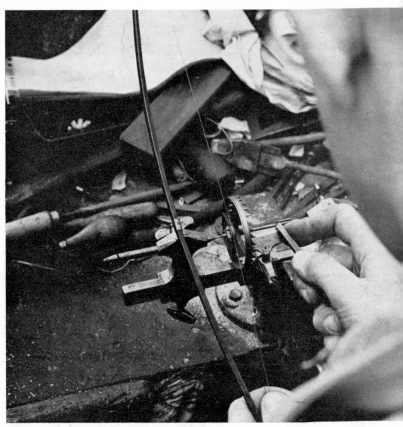

PLATE 127 Using the Turns. The drive is by the bow. The bowstring grips the ferrule which is in two pieces. The tool is held in the right hand supported by the adjustable tool rest and has to be slightly moved out of the way on the return stroke when no cutting takes place. The piece being worked on is the scapewheel of a 'Comtoise.'

PLATE 128 Throw for turning the arbor of a turret clock. The arbor is mounted between the centres and is rotated by the handwheel through the belt. The drive is transmitted to the arbor by the oval shaped 'carrier' which is secured by a screw. The tool is supported by the adjustable tool rest supported by the middle of the bed.

PLATE 129 A Clockmaker's Throw. The left hand operates
the wheel and the right hand holds the tool on the tool rest
while the work rotates between the centres. The horizontal
bar behind the centres is a steady for supporting long arbors
while they are being turned. In the background is a wooden
pattern for casting a wheel for a turret clock.

PLATE 130 A clockmaker's screwcutter. The size and number of threads cut by the tool which is carried at the left can be varied by altering the setting by means of the wing nuts on the right. One turn of the handle produces one turn of the thread.

Crown Copyright. Science Museum, London.

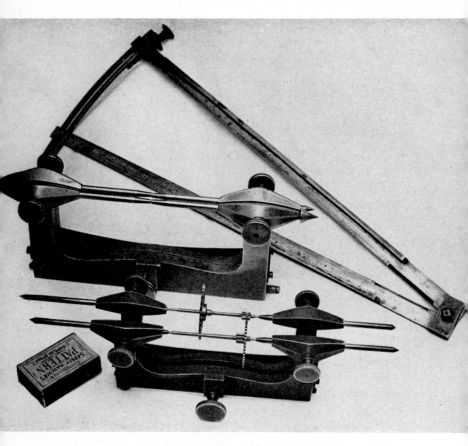

PLATE 131   A pair of Depthing tools. Adjacent wheels and pinions are mounted in the tool and their distance apart can be varied by means of the screw in the centre at the bottom. When the wheel and pinion mesh comfortably, the points at the end are used as dividers to mark the plate for drilling.

II

WHERE TO SEE CLOCKS

The London collector is probably the most fortunate of clock lovers as there are more opportunities of seeing clocks there than in any other city or town in the country.

Pride of place must be given to the British Museum, for in that building is not only housed the Museum's own collection, but the collection of the late C. A. Ilbert was added in 1958, and is probably one of the most comprehensive in the world. Ilbert was a civil engineer who made a hobby of horology, and his outlook was broad enough to include all types of clocks and watches in his collection. Even the plain Waterbury watch was included. A selection from both the museum collection and the Ilbert collection is on show in the gallery, and in addition there is a student's room where the rest of the collections are stored and specimens can be examined closely. It is necessary to make an appointment to use this room, but it is well worth the trouble. The collection includes an early Long Case with short pendulum and verge escapement, Gothic wall clocks with balances and a 'Cross Beat'.

The collection of the Clockmakers' Company is housed in the Guildhall (entrance in Basinghall Street). Although this consists mostly of watches, clocks are also included. An early Tompion Long Case clock is displayed, together with a Harrison Long Case with wooden wheels. An astronomical clock by Samuel Watson is of interest, and also a Coster type pendulum clock by Pieter Visbach.

The Science Museum

The Science Museum collection has been assembled mainly from the technical point of view, but this should not deter the non-technically minded clock lover from visiting it. The movement of the Wells Cathedral clock is exhibited going, although it still retains the pendulum and anchor escapement it acquired in the seventeenth century, and has not been re-converted to verge escapement with foliot balance. The main interest in this clock is therefore its chiming mechanism, which was probably made only three years after the first quarter chimes were installed at Rouen. It is interesting to watch the clock strike at the hour and see how the quarter mechanism lets off the hour strike after the four quarters have been sounded.

The collection also contains another ancient turret clock with a wooden frame from Worcestershire, and of course, the famous Dover Castle clock which still retains its foliot balance and verge escapement. Early weight-driven wall clocks are shown, and the collection also includes English Lantern and Bracket clocks. There are a number of models of escapements, and two of the clocks described by Huygens in his writings have been reconstructed and are shown going. A Coster type clock and a 'Stoelklok' also feature in the collection. The more primitive time measurers such as Astrolabes, Nocturnals, Sand Glasses, Clepsydrae, Marked Candles etc., are also represented.

The Victoria and Albert Museum just across the road from the Science Museum has a collection of clocks and watches assembled more from the artistic point of view. Early table clocks are included together with Gothic wall clocks, and various English Bracket and Long Case Clocks are displayed in the galleries containing furniture.

The London Museum in Kensington Palace includes a few clocks in its collection with the stress naturally on London makers. The Wallace Collection in Manchester Square probably has assembled the largest number of French clocks in London. Many different types are included and give the student of French horology quite a lot of varied material.

The last collection in London to be noted is that of the National Maritime Museum at Greenwich, which includes the original Greenwich Observatory. The stress here is naturally on Navigation, and the collection is less likely to be of interest to the clock lover than the other London collections, but it should be visited if only to see the exquisite workmanship of Harrison's four marine timekeepers and the reproductions of the Tompion clocks in the Observatory itself. One of the original movements for these clocks is preserved in the British Museum collection, but now has a one-second pendulum. As fitted at Greenwich, the pendulums were thirteen feet long beating once in two seconds, and hung above the movements. A small window allows the visitor to see how small the amplitude of the swing of these pendulums was. The Maritime Museum also displays some Astrolabes and Nocturnals.

Clocks can be studied not only at museums but also in Auction Rooms and Antique shops. Christie's, Sotheby's, Knight, Frank and Rutley and Phillips, Son and Neale among others often include clocks in their sales, and the smaller Auction Rooms are not to be neglected either. Antique shops in London are mostly in the Kensington-Chelsea area, but there are also one or two in Bond Street and other parts of the West End. A new group is springing up near Islington Green, and the Caledonian Market near Tower Bridge and the Portobello Road are both worth a visit.

Outside London, collections are spread over a wide area, though none are as large as those in the metropolis. Most local museums contain one or two specimens of clocks from the surrounding district, and several have made a feature of clock collections. South and South East of London are the museums at Rochester and at Maidstone, with examples of clocks by Kentish makers, and the museums of the Sussex Archeological Society at Barbican House, Lewes, Anne of Cleves' House, Southover, and the Priests' House, West Hoathly. Barbican House has examples of thirty-hour Long Case clocks and a Lantern clock, together with an American wall clock of the type that was imported into this country in large numbers. Anne of Cleves' House has an eight-day Long Case clock, an 'Act of Parliament Clock', a Stoelklok and a turret clock from Wittersham Church in Kent of 1826. At West Hoathly are one or two domestic clocks and some parts removed from the local church clock when it was repaired. The town of Lewes contains one or two antique shops where clocks are often displayed.

Nearby Brighton is famous for its 'Lanes' where antique shops abound, but the local museum only contains one Long Case clock by a local maker. Hove Museum has a large collection of watches, but very few clocks. At Preston Manor just outside Brighton the collection includes a marquetry Long Case clock and a French travelling clock.

Going west of London, we have Windsor Castle where the State Apartments are frequently open to the public, and here are found both English and French clocks of the eighteenth and nineteenth centuries, together with a late seventeenth-century turret clock in the Curfew Tower.

The Ashmolean Museum at Oxford contains some interesting specimens of Table clocks etc. as does the Museum of the History of Science in that city, while further west over the Welsh border at Cardiff is the National Museum of Wales with its branch at St Fagan's Castle and the collection of Welsh clocks and watches that has inspired a book on the subject which has run into more than one edition. Snowshill Manor, a National Trust property near Broadway, displays some clocks in its collection.

In Ireland, the National Museum in Dublin contains some specimens of clocks by Irish makers together with a Stoelklok.

North and East of London, the Fitzwilliam Museum at Cambridge houses among other things some very early Long Case clocks. The Gershom Parkington Collection at Bury St Edmunds contains specimens by most of the great London makers together with some foreign pieces, and the museum at Colchester possesses a number of clocks by Colchester makers. The museum at Leicester contains a collection of clocks that are described in a small brochure, and farther north at Lincoln is the Usher Art Gallery. The Castle Museum at York includes a Clockmaker's shop in its street of old-time shops.

The City Museum at Liverpool is building up a collection of Liverpool clocks, and also possesses the remains of the large clock constructed by Jacob Lovelace of Exeter which was severely damaged in an air raid in 1941.

Turret clocks are usually inaccessible to the public but several early specimens are on view in various parts of England. If one wishes to see them on the Continent it is usually necessary to visit a museum. In the South East of England there are old turret clocks preserved at Folkestone Museum (out of the Parish church) and at Canterbury Westgate (from Godinton church). Rye church in Sussex boasts of having the oldest public clock in England still running with its original movement. While the frame of this clock is old, many parts have been renewed and it has been converted to electric winding in recent years. The long pendulum which hangs down inside the church is a later addition, as is also the quarter chiming mechanism. During the summer, the tower is usually open to visitors.

The turret clock at the Anne of Cleves Museum, Lewes, has already been mentioned, and the former clock from Shoreham Parish Church is also preserved in the Marlipins Museum in that town. The clock at Groombridge, Kent can easily be seen from the floor of the church, and is one of the few left still possessing only one hand.

In the West of England are the four clocks formerly attributed to Peter Lightfoot – Wells, Wimborne, Ottery St Mary and Exeter. The original Wells movement is preserved in the Science Museum, the movement now actuating the dial dating from 1886. The Wimborne clock is now provided with a movement dating from about 1740, and the Exeter clock also has a modern movement, the old one being displayed in the North Transept of the Cathedral. The old Exeter movement is such a mixture of styles as to be of little antiquarian value, and only the

Ottery St Mary movement can be considered reasonably complete. This clock underwent restoration in the early years of the present century and some alterations were made to it, but it can be considered a good representation of a quarter chiming clock with an astronomical dial, and represents the probable appearance of the earlier Wimborne and Exeter movements.

Salisbury, of course, gives the best impression of an old turret clock. The traditional date of its making is 1386, and the restoration it received in 1956 can probably be accepted as restoring it to its original appearance.

Cotehele House in Cornwall possesses an old turret clock, but the layout of its movement is unconventional. It has its verge escapement at the bottom with the other wheels arranged above it. It never possessed a dial. The Cotehele clock is made of iron, but resembles a number of clocks with wooden frames that exist in the Midlands.

Finally in the West of England we have the clock of Sherborne Abbey, which dates from the eighteenth century and resembles the present movement at Wimborne.

Another opportunity of seeing old clocks is provided by visiting the Stately Homes now open to the public. The number of clocks is not large, but the pieces displayed are sometimes of high quality. Country hotels and tea rooms often include old clocks in their furnishings although they are less frequently met with than they were a few years ago.

The opportunity of seeing old clocks does not present itself so frequently on the Continent as in England. Collections are often at places a great distance apart and it is not always convenient to travel from one to the other.

The visitor to France is best catered for in Paris. Here are the collections of the Conservatoire des Arts et Métiers, The Louvre, the Musée des Arts Décoratifs, and the Petit Palais. Just outside Paris is the Palace of Versailles which contains a number of clocks that cannot be very easily seen, as they form part of the furnishings of rooms which are roped off from the public.

The Musée des Beaux Arts at Besançon contains part of the Gélis collection, the bulk of which is housed in the Musée Paul Dupuy at Toulouse. The latter collection is the largest outside Paris. The Musée Calvet at Avignon has a small collection of clocks and watches, including two iron wall clocks of the fifteenth century from Burgundy. The Chateau of Pau contains a number of Religieuses.

Strasbourg Cathedral contains the world-famous astronomical clock, but it is now actuated by a movement of 1842. The older movement by the brothers Habrecht is housed in the local museum, and is well worth a visit. The great clock at Rouen is not usually on view to the public.

Holland is fortunate in the number of its horological exhibits, and as it is not a very large country travelling is made easier. The main clock museum is at Utrecht and contains specimens of the various types of Dutch clock together with examples from other parts of Europe, as well as the clock from the St Jacobstoren at The Hague dated 1542, which possesses a carillon together with its verge escapement and foliot and is a turret clock in the real grand manner. Just outside The Hague is Hofwijk, the country house of Christiaan Huygens, where is preserved the pendulum and escapement of the church clock at Scheveningen, the first turret clock to be fitted with a pendulum. The house also possesses a Coster type clock but nothing else of horological interest.

A few miles to the north at Leiden is the Dutch Science Museum. The collection here includes the earliest known pendulum clock by Coster and one or two other early clocks, together with a number of models of timekeepers devised by Huygens after he had invented the pendulum clock. His aim was to produce a timekeeper for recording accurate time at sea, but none of these ideas were so effective as his original one, which unfortunately was only suitable for clocks on land. The Frans Hals Museum at Haarlem contains one or two Dutch clocks including a Zaanse clock and a Bracket clock. The Rijksmuseum at Amsterdam possesses one or two clocks in its furniture section. In the north of the country, the Friese Museum at Leeuwarden includes some typical Frisian clocks in its collection and also possesses a set of clockmaker's tools and two small turret clocks. These are kept at a country house belonging to the museum some miles out of Leeuwarden, but are not generally on show to the public. The museum at Groningen includes some representative types of Dutch clock, including the type of Stoelklok associated with that province. In the centre of Holland is the Open Air Museum at Arnhem which shows the type of house associated with each province, and includes several clocks among the furnishings.

Germany possesses a number of museums containing clocks, but it must not be forgotten that the distances in this country are very great and it is difficult to visit all museums in a short space of time. In the north we have Wuppertal where the collection includes clocks of many types from all parts of the world, and Kassel where are found the observatory clocks by Jost Burgi and a number of other interesting clocks. Going further south to the Black Forest, we find museums at Triberg and Fürtwangen. These museums specialise on Black Forest products. Fürtwangen has the larger collection and also includes representatives from other countries. A period workshop has been reconstructed and shows a selection of old tools as well as the clocks that they were used to produce.

Schwenningen is a manufacturing town mainly engaged in the horological industry and two of the factories concerned, Mauthe and Kienzle, have assembled collections of old clocks and watches. These museums are only open at certain times. The town museum here also has some clocks in its collection. The collection of the Landesgewerbeamt at Stuttgart contains a number of Black Forest and other types of clock, but these are not generally on show to the public. The collection of the Germanisches Museum at Nuremburg contains the controversial spring clock of Philip the Good together with other early clocks, and the museum at Augsburg contains some of the horological specialities of that city. The Deutsches Museum at Munich is the German equivalent of the

English Science Museum, and also has a clockmaker's workshop fitted up, together with a number of examples of German and other clocks. The Bayerisches Museum in the same city also has some clocks in its collection. At Bamberg in Eastern Bavaria are to be found some examples of the type of Bracket clock associated with that district. The castle at Wurzburg is not rich in horological specimens but contains what is probably the oldest wall clock in the world. It has only two wheels in its train and is, of course, controlled by a verge escapement and foliot. Before we leave Germany, it should be mentioned that the Junghans factory hopes to establish a museum of its own products in the near future.

The rest of Europe is not rich in collections. In Switzerland are found the Kirschgarten Museum in Basel which houses some wall clocks including some examples by the famous Liechti family of Winterthur, and some table clocks of both early and late types. Other Swiss museums of horology are found at Le Locle, La Chaux de Fonds and Neuchatel. Austria possesses its own clock museum in Vienna where the maximum number of Austrian clocks can be seen and the Museum of Tyrolean Folklore at Innsbruck contains early wall clocks with the short pendulum arranged to swing in front of the dial. One of these has its weights boxed in, virtually turning it into a Long Case clock. At Bad Ischl, the Léhar Villa contains a Bracket clock, a very tiny Vienna Regulator and a Long Case clock with a short pendulum swinging before the dial. The Emperor's country villa in the same town possesses several more examples of Austrian clockmaking.

In Belgium there are the Museés Royaux D'Art et D'Histoire in Brussels and the Vleeschal at Antwerp. In the Musée de la Vie Wallon at Liége are some sundials, but no clocks.

Denmark has museums at Aarhus, 'Den Gamle By', Copen-hagen, The National Museum, and Sorgenfrei near the latter city, where the collections include clocks. In Sweden are the Stads-museum in Stockholm, the Nordliches museum in the same city and the Upplands museum in Uppsala.

Far to the South in Italy is the Leonardo da Vinci collection in

Milan. It is most appropriate that clocks should be displayed here, as Milan claims to have had the first public clock in the world.

In Spain there is the Royal Palace in Madrid, while in San Sebastian there is a large and varied collection at the Hotel Biarritz which has been acquired by the proprietress over many years.

In the United States of America there is the Bristol Clock Museum at Bristol, Connecticut, which specialises in American clocks, while in New York is the Metropolitan Museum of Art which has a collection from a wider area. The James Arthur Collection at New York University contains both American and European clocks. The Smithsonian Institution in Washington D.C. has a horological section, and shows a replica of the Dondi clock of 1348–64 which was made in London in 1960 from the information contained in Dondi's writings.

The Illinois State Museum at Springfield houses the Hunter Collection of clocks which includes both American and European clocks, the latter embracing products of England, Holland, Germany and France. The Dutch representatives are a Long Case clock, a Staartklok and a Zaanse clock with a separate quarter-hour dial. Farther west are the Hagans Clock Manor Museum at Denver, Colorado with a very large collection of horological items, and the California Academy of Science at San Francisco which houses the Dr William Barclay Stevens Collection.

As in Britain, many museums in the United States contain horological exhibits which form only a small portion of the museum's total collection, so it is a good practice for the clock enthusiast in both countries to make a routine visit to the museum in every town he passes through. One can never tell what objects of interest may be discovered by this method.

12

GLOSSARY OF
TECHNICAL TERMS

Act of Parliament Clock A wall clock with a weight-driven movement like that of an English Long Case clock and a dial of up to
3 ft wide. These clocks generally have no glass over the dial and
are often lacquered, having the dial black with gold figures.

In 1797 the British Government imposed a tax on clocks and
watches and the legend has it that this type of clock was installed
in inns to enable the public to know the time without needing to
possess watches and clocks of their own and pay the tax. The type
was, however, well known before 1797 and continued to be made
until the nineteenth century.

Arbor The axle on which a wheel is mounted. The term is derived from the Latin for 'tree' cf 'Axletree'. Each arbor in a clock
usually bears a wheel and a pinion. A pinion is a wheel of less than
20 teeth and its teeth are usually known as 'leaves'.

Arc A portion of the circumference of a circle. The term is
usually employed in clockwork to describe the path of a pendulum
bob.

Amplitude The distance through which a pendulum bob moves as
it swings.

Balance A mass of metal, usually in the form of a wheel or bar
that oscillates about its centre and controls the speed at which a
clock or watch runs. It operates independently of gravity, and in
the early days it had no definite period of its own, making the

going of the clock or watch very erratic. The invention of the balance spring about 1675 enabled the balance to be used for exact timekeeping, and when the effect of variations in temperature had been overcome in the eighteenth century it was possible to construct timekeepers to give accurate time on long voyages at sea.

Bob　The weight at the end of the pendulum. The earliest bobs were pear-shaped, later ones were lens-shaped and bobs for precision clocks or turret clocks are now usually made cylindrical.

Boulle or Buhl　Charles André Boulle was born in Paris in 1642 and became celebrated as a chaser and inlayer. In 1672 he was allocated rooms in the Louvre and his special inlay work using brass and tortoiseshell became very fashionable. It lasted well into the eighteenth century.

Bush　A small circular piece of metal with a hole through the centre to provide a bearing for a pivot to run in. Black Forest clocks used to have brass bushes inserted in wooden plates, and bushes were also used when the material of a metal plate was too soft to stand normal wear. When pivot holes wear large, a bush is inserted to keep the arbor in its correct position. The bushes are hammered into the plates after the holes have been opened up nearly to the size of the bush, and then the pivot holes in the bushes are also opened up to their correct sizes with a broach. (See Chapter 10).

Chapter Ring　The ring on which the figures are engraved or painted, usually having a contrasting finish to the surface of the dial.

Clickwork　A device which allows a fusee or barrel arbor to turn freely in one direction to permit winding, but keeps it firmly in connection with the train when rotating in the opposite direction. On a clock with pull up wind, this can consist in its simplest form of a spring bearing against the spokes of the great wheel.

Crutch　A lever fixed to the pallet arbor or verge with a forked or looped end that embraces the pendulum and constrains the arbor or verge to move in time with the latter. It also serves to transmit the impulse from the escapement to the pendulum. The crutch was used in the earliest pendulum clock of Coster, but was abandoned by English makers in the verge-escapement clocks that they developed from the Coster design.

Cycloid A curve generated by a point on the circumference of a circle which is rolling along a straight line.

Escapement The escapement is the device which allows the last wheel of the train to advance only in a series of short movements. It is restrained by the balance or pendulum but provides the necessary impulses to keep the latter running.

The earliest escapement for clocks and watches was the verge. The verge itself is a long thin rod bearing a pallet near each end, the two pallets being at an angle of about 90° from each other. The wheel which acts on the pallets is known as the crown wheel on account of its shape. The driving force causes the crown wheel to rotate until one of its teeth intercepts one of the pallets and attempts to push it out of the way. This causes the verge to rotate, and by the time the tooth has pushed the pallet out of its path the other pallet has advanced in front of a tooth on the opposite side of the wheel and the process begins again. The balance itself is attached directly to the verge, or in the case of a pendulum clock the pendulum may be either attached to the verge or connected to it by means of a crutch.

The Cross Beat escapement acted on the same principle except that the wheel was flat and the verge duplicated, each part bearing one pallet only. Each of the verges had a foliot which was lighter than those usually fitted, and a certain amount of springiness in the arms helped to bring the system nearer to having a regular period of its own.

Next in order of time came Galileo's escapement. Here the wheel had saw-shaped teeth and an equal number of pins projecting from the rim parallel to the arbor. It was normally held stationary by a catch which prevented it from turning by obstructing one of the teeth. The pendulum swung from pivots, and on the same centre were two arms very like long commas which moved in unison with the pendulum. As the pendulum swung towards the wheel, one of the arms would touch one of the pins and cause the wheel to recoil a little, while the other arm would raise the catch. As the pendulum completed its swing and began to return, the force of the mainspring caused the wheel to rotate and give impulse to the pendulum by means of the pin in contact with the arm. By the time the pin had escaped from the arm, the other arm had sunk low enough to let the catch return to its normal position, and it held the wheel stationary again by impeding the next tooth. This escapement never came into general use but the basic idea was used for the escapement applied to marine chronometers a century and a half later.

The anchor escapement developed for the Long Case clock was very simple. The fronts of the teeth were concave curves which were arrested by each pallet and gave impulses as they escaped while the pallets moved under the influence of the pendulum. As a tooth escaped from one pallet, another one on the other side of the wheel would be caught by the pallet there and so on.

All the escapements mentioned so far were recoil escapements, that is those where the wheel moved back a little before it advanced. George Graham, Tompion's successor, devised an escapement with a dead beat in which the wheel advanced by a series of jumps without recoiling. This gave more satisfactory results and was used for clocks where a very high standard of accuracy was required. Only the extreme points of the teeth were used, and the faces of the pallets were circles struck from the same centre as that of the pallet arbor.

French makers favoured a version of this escapement where the teeth were replaced by pins fastened to the rim of the wheel parallel to the arbor, and the Graham type pallets were placed much closer together. Each pin would act on each pallet in turn.
Foliot   A bar balance used in conjunction with the verge escapement. A number of grooves are cut near the end to provide alternative positions for the small weights which can be moved nearer to or further from the centre for regulation purposes.

Glossary

Floating Balance   A balance supported by the balance spring in order to eliminate the friction between the normal pivots of a staff and their bearings.

Going barrel   A drum with teeth round its edge containing a spring, which conveys the power of the spring directly to the train without the intervention of a fusee.

Gothic Clock   A name often applied to the mediaeval weight-driven wall clocks.

Jig   A piece of metal accurately cut to shape, which guides the worker in producing parts of the same shape.

Lifting Piece   Name applied to the levers which release and lock the striking work.

Locking Plate   A disc with notches cut in its edge at increasing intervals to control the number of blows struck.

Marquetry   Several thin sheets of wood of different colours are fixed together and a pattern is cut through them with a fretsaw. The sheets are then separated and the cut out portions rearranged in a panel of different colour similarly to a jigsaw puzzle. Popular on English clocks till the early eighteenth century.

Morbier   Name sometimes applied to Comtoise clocks derived from the district where they were made.

Motion Work   The arrangement of gearing whereby the hour hand rotates once in twelve hours while the minute hand rotates once every hour. The minute hand is fastened to a tube which fits very closely on the centre arbor and is known as the cannon pinion. This carries a pinion or small wheel at the rear end and drives a wheel which rides on a stud, known as the minute wheel. The latter has a pinion at its centre which drives a wheel fastened to the tube which carries the hour hand, which is placed over the tube carrying the minute hand and rides loosely on it. The cannon pinion can either be fitted friction tight on to the centre arbor or in English clocks it is normally holding a small flat spring in compression which makes it rotate with the centre arbor as the clock goes, but leaves it sufficiently free for the hands to be adjusted as necessary. Sometimes on English thirty-hour clocks and Dutch and Black Forest wall clocks the drive is through the minute wheel which is then mounted on an extension of the main arbor. The friction drive is then incorporated at this point and the tubes carrying the hands are carried on a long stud which projects through the dial. The motion work is arranged to release the striking so that the clock strikes when the minute hand is exactly on XII.

Movement   The wheels and other moving parts of the clock together with the frame.

OG   A term applied to the cases of certain American clocks on account of the front moulding being shaped in an ogee curve.

Ormolu   Literally 'Ground Gold'. A finish applied to the cases of French clocks giving a rough gold surface which is capable of being polished afterwards.

Pallet   A piece of metal which interrupts the motion of the escapement wheel ('Scape wheel') and is controlled by the balance or pendulum, thereby making the clock run to time.

Parquetry   Veneer applied to a clock case in the form of a number of pieces of regular shape.

Pinion   Technically a wheel with less than twenty teeth. The teeth of a pinion are usually called 'leaves'. Pinions are of two sorts, 'solid' which are either cut from a solid piece of metal or made from pinion wire, and 'lantern' which are built up out of two discs connected by wires forming a sort of cage.

Pivot   The narrow end of an arbor which runs in the hole in the frame and forms the bearing surface.

Platform   A small metal plate usually applied to the movement of a carriage clock containing a watch escapement and balance. The work is finer than that of the clock movement generally.

Rating Nut   The nut below the bob of the pendulum which allows adjustments in the clock's rate to be made by raising or lowering the bob.

Scapewheel   The last wheel in the going train of a clock which operates the escapement.

Spandrel   The decorative corner piece of the dial.

Glossary

Stackfreed   A friction brake used in the sixteenth-century German table clocks to even up the power of the mainspring without the use of a fusee.

Staff   The arbor of a balance, usually associated with the Lever escapement.

Ting Tang   A quarter chime where two notes are sounded for each past quarter, the second being lower than the first.

Train   A set of wheels and pinions forming a continuous drive.

Turret Clock   A large clock placed in a tower.

Vernis Martin   A form of decoration invented by Robert Martin of Paris in the early eighteenth century, in imitation of oriental lacquer.

13

BIBLIOGRAPHY

VON BASSERMANN JORDAN-BERTELE, *The Book of Old Clocks and Watches* (Allen and Unwin, 1964)

F. J. BRITTEN, *Old Clocks and Watches and their Makers* (Batsford, 1933) (Revised edition, Spon, 1956)

The Watch and Clockmakers Handbook Dictionary and Guide (Spon, 1955)

P. G. DAWSON, *The Design of English Domestic Clocks 1660–1700* (Antiquarian Horological Society, 1956)

G. F. C. GORDON, *Clockmaking Past and Present* (Technical Press, 1946)

ERNEST L. EDWARDS, *Weight Driven Chamber Clocks of the Middle Ages and the Renaissance* (Sherrat, 1956)

The Grandfather Clock (Sherrat, 1956)

H. ALAN LLOYD, *Old Clocks* (Ernest Benn, 1958)

The Collector's Dictionary of Clocks (Country Life, 1964)

Some Outstanding Clocks over Seven Hundred Years (Leonard Hill, 1958)

J. DRUMMOND ROBERTSON, *The Evolution of Clockwork* (Cassell, 1931) This work contains an extensive bibliography as well as much valuable material on Japanese clocks.

R. W. SYMONDS, *A History of English Clocks* (Penguin, 1947)
Thomas Tompion : his Life and Work (Batsford, 1951)

TARDY, *La Pendule Française* (1965)

UNGERER, *Les Horloges Astronomiques et Monumentales* (Strasbourg, 1931)

F. A. B. WARD, *The Science Museum Handbook on Time Measurement* (Parts I and II)

Bibliography

Collectors' Pieces: Clocks and Watches (Antiquarian Horological Society, 1964)

PERIODICALS

Chronos (Holland) fortnightly.
La Clessidra (Italy) monthly.
Heft der Freunde Alter Uhren (Germany) annually.
Bulletin of the National Association of Watch and Clock Collectors (U.S.A.) Published five times a year.
Antiquarian Horology (Britain) quarterly.
Horological Journal (Britain) monthly.

Acknowledgements

The author and publishers wish to thank the following for permission to reproduce photographs:

Trustees of British Museum: Plates 1, 2, 5, 6, 7, 8, 9, 10, 25, 27, 28, 37, 42a, 46, 55, 114, 117; Science Museum, London (Crown copyright): 4, 11, 26, 39, 49, 53, 81, 126, 130; Wallace Collection (Reproduction by permission of the Trustees of the Wallace Collection): 83, 84, 85, 86, 87, 88, 89, 90, 91, 93; Uhrenmuseum der Stadt Wien: 82, 110, 111, 112, 113; National Museum, Helsinki: 98, 99, 100(b), 101, 102, 103, 104, 105, 106, 107, 120; Archivio Fotografico dei Civici Musei, Milan: 108(b); Mr J. Zeeman, Netherlands Clock Museum, Utrecht: 3, 41, 47, 48, 50, 51, 52, 54, 58, 59, 79, 80; Sammut Collection, St Venera, Malta: 115, 116; Norsk Folkemuseum, Oslo: 109; National Museum, Stockholm: 97, 100(a), 108(a); Mr T. P. Camerer Cuss: 60, 61, 63, 65, 66, 67, 68, 69, 70; Mr C. E. Clear: 78; Mr P. G. Dawson: 29, 31, 32, 33, 36, 38, 42(b), 43(c), 44, 94, 95, 119; Mr S. G. Edgcombe: 12, 13, 14, 15, 16, 17, 18, 19, 20, 21, 22, 23, 24, 30, 34, 35, 40, 43, 97, 122; Dr P. R. Latcham: Frontispiece; Messrs. A. A. Osborne & Son: 123, 125, 127, 128, 129, 131, 132; Messrs. Sotheby & Co.: 45.

INDEX

Index